PERMACULTURE & REGENERATIVE
LANDSCAPE DESIGN
WORKBOOK

By Stephanie Lindhardt

ASSESSMENTS & KEY CONCEPTS

LEARN KEY CONCEPTS WITH WORKSHEETS THAT HELP YOU ASSESS YOUR PROPERTY'S POTENTIAL

SPECIAL NOTE TO THE READER:
THIS IS NOT A PERMACULTURE COURSE

Permaculture is an organized framework for understanding how to care for the earth by working with natural ecosystems. There are countless books, courses, and articles that explain who developed it, what it encompasses, and how it works. This workbook is not about those histories or theories, it's about helping you directly apply permaculture and regenerative agriculture concepts to your land in practical, meaningful ways.

You do not need any prior knowledge of permaculture or regenerative agriculture to use this book. However, if you're interested in exploring the 12 principles and 3 ethics that guide permaculture design and management, I recommend visiting *www.permacultureprinciples.com*

In order to condense as much information as possible into a practical workbook, concepts are given throughout the book in a brief and succinct way without much exploration into what each concept could entail. I encourage you to research concepts further if you find anything particularly interesting or relevant to your land.

TABLE OF CONTENTS

NOTE ON INIGENOUS INFLUENCE WITH GREAT RESPECT..6

HOW TO USE THIS WORKBOOK..8

WEBSITES YOU NEED TO KNOW 10

SECTION 1: INTEGRAL FOUNDATIONAL DESIGN ASPECTS...11

    LEGAL CONCEPTS..12

    CREATING A BASE MAP..15

    PROPERTY QUESTIONNAIRE..18

    LEGALITIES..22

    HAZARD ASSESSMENT..28

    LOCAL RESOURCES..38

SECTION 2: SYSTEMS THINKING..40

    SYSTEMS THINKING CONCEPTS..41

    SECTOR MAPS..43

    ZONES OF USE..46

    CLOSED LOOP ASSESSMENT..49

    WHAT GOES WHERE..51

SECTION 3: METEOROLOGY ASSESSMENT...52

    METEOROLOGY CONCEPTS..53

    PROPERTY CLIMATE DATA..56

    CALCULATING THE RAINWATER BUDGET..61

    SOLAR ASPECT MAP..62

SECTION 4: GEOLOGY ASSESSMENT..66

    GEOLOGY CONCEPTS..67

    SOIL ANALYSIS..69

    SOIL ACIDITY TEST..71

    GEOLOGY ANALYSIS..72

SECTION 5: TOPOGRAPHY ASSESSMENT...73

    TOPOGRAPHY CONCEPTS..74

    WATER FLOW MAP..77

    TOPOGRAPHY ANALYSIS..80

SECTION 6: ECOLOGY ASSESSMENT..81

    ECOLOGY CONCEPTS..82

    ECOLOGICAL ANALYSIS..85

    FOREST LAYERS..88

    TREE GUILD BUILDER..90

    EDGE EFFECT ANALYSIS..91

    ENDANGERED PLANT ANALYSIS..93

    PREDATOR PEST ANALYSIS..94

    AT RISK & ENDANGERED SPECIES ANALYSIS..97

    KEYSTONE WILDLIFE SPECIES ANALYSIS..100

SECTION 7: BIOLOGY ASSESSMENT...102

    BIOLOGY CONCEPTS..103

    BIOLOGY ANALYSIS..105

DESIGN ORDER & CONCLUSION...106

NOTE ON INDIGENOUS INFLUENCE WITH GREAT RESPECT TOWARDS THEIR LAND MANAGEMENT PRACTICES:

Indigenous communities around the world have developed sophisticated knowledge systems over generations, harmonizing with nature to ensure the long-term fertility, resilience, and productivity of their lands. The various modern movements of permaculture, agroforestry, and regenerative agriculture are inspired by these ancient traditions that came from indigenous wisdom and all credit goes to them with great respect and gratitude. Here's how some indigenous land management practices have influenced these formalized practices and how we can pay respect to indigenous communities when designing:

Observation and Mimicry: Indigenous communities have historically been keen observers of nature, understanding intricate relationships between different elements in an ecosystem. Indigenous practices of observing and learning from the land is a foundational construct to all the various formalized practices such as permaculture and regenerative agriculture.

Biodiversity Conservation: Many indigenous cultures value biodiversity and practice polyculture (growing multiple crops in the same space). Indigenous peoples' deep understanding of plant and animal interactions heavily inspired various key constructs within permaculture, agroforestry, aquaculture, and the other various regenerative agriculture movements.

Regenerative Agriculture: Indigenous land management often involves regenerative agricultural practices such as agroforestry, cover cropping, and crop rotation. These techniques have been incorporated into many of the formalized practices by focusing on soil health and regeneration without depleting natural resources.

Water Management: Indigenous communities have developed intricate systems for managing water, including rainwater harvesting and aquaculture. These techniques have been incorporated into many modern projects that are regreening the desert, increasing drought resilience, and mitigating flood damage.

There is a lot of knowledge to be gained from learning about indigenous land caretaking practices in your area. Research your local tribes & their customs to respectfully learn more.

CULTURAL RESPECT & ETHICAL CONSIDERATIONS WHEN DESIGNING

When designing with permaculture & regenerative principles, it is crucial to acknowledge and respect your local indigenous tribes and the knowledge whereby most of these practices were formalized. This can be done by:

- Acknowledge what indigenous land you live on and which tribes historically cared for that land. There is much to learn from the traditional ways of the indigenous land caretakers in your area.You can find out the historical indigenous tribes and boundaries of your area here: https://native-land.ca/

- Give thanks and credit where it's due and educate others about the influence of indigenous knowledge and sustainable practices.

- Engage with indigenous communities. Learn from their experiences, and listen to their stories. Respect their traditional knowledge and wisdom.

- Collaborate with indigenous communities by buying from Native and Tribal companies. Purchase educational books written by First Nations authors, handmade tools, seeds, or other market goods from indigenous peoples.

- Avoid cultural appropriation and commodification of indigenous practices. The credit doesn't belong to those who organize the information, it belongs to those who created the information. Respect the sacredness of their rituals, ceremonies, land, and traditions.

HOW TO USE THIS WORKBOOK

This workbook is designed to guide you through both learning permaculture and regenerative concepts while gathering the data needed for your land assessment and design. Rather than keeping theory separate from practice, the workbook blends the two, allowing you to build knowledge while directly applying it to your own property.

This workbook is organized into seven main sectors relevant to land design and management. They are:

- **Legalities - Ensuring all land and water laws are being followed**
- **Systems Thinking- integrating all elements into a cohesive, regenerative design**
- **Meteoroglical - Climate, weather patterns, and atmospheric conditions**
- **Topographical - Slope, elevation, and landform features**
- **Biological - Living systems within the soil & property**
- **Geological - Land and soil composition in relation to its surroundings**
- **Ecological - Natural cycles, interactions, and resource flows**

Each sector begins with a simple introduction to key concepts, like a mini dictionary, that defines and explains the terms you'll need to understand within that sector. Following this, you'll find worksheets that help you record and analyze the actual conditions of your property relating to that sector's analysis.

These exercises are designed so that by the time you finish, you will have both a solid grasp of permaculture & regenerative principles, a detailed set of data to use in your design process, and ultimately a design that is aligned with Mother Nature that you can feel confident about. These pages can be reused again and again by individuals and professionals alike to organize projects and ensure a cohesive and thorough design process.

To get the most out of this workbook:

1. Begin this workbook and work through each sector in order, reading the concept overviews first.

2. Use the worksheets in each sector to collect information specific to your property.

3. Treat your completed workbook as both a learning tool and a working field guide, something you can return to and update as your property evolves.

By the end, you'll not only understand the foundations of permaculture and regenerative design analysis, but you'll also have a customized land assessment portfolio that can guide your decisions, help you design for resilience, and keep your landscape ecologically balanced for years to come.

WEBSITES YOU NEED TO KNOW

Use this checklist to ensure you've researched the key legal considerations for your property design. Refer to this list as necessary during the research, design, & installation of your project.

PRO RESEARCH TIP
To find your regional websites google: [CountyCity/State] then list the department. e.g. California FEMA flood map

COMMUNITY & NEIGHBORHOOD REGULATORY SECTORS
- Homeowners Association (HOA) - Landscaping/building regulations, animal restrictions
- Neighborhood Community Codes & Restrictions (CC&R's) - Landscape/building regulations, special design, historic distric regulations

CITY & COUNTY GOVERNMENT WEBSITES
- County Recorder or Assessor's Office - Property boundaries, parcel maps, ownership details, tax info
- County Recorder/Clerk's Office - Deeds, easements, water rights records
- City Planning & Zoning Department -Zoning codes, permitted land uses, setbacks, outbuilding/animal restrictions
- City Ordinances Database - A website hosting the legal municiple codes regarding zoning, land use, and restrictions.
- City Building & Permits Department - Buildings, septic, grading, and permits

WATER & IRRIGATION RESOURCES
- State Water Resources Department - Water rights, usage, restrictions
- County Water Conservation Districts - Water rights, usage, restrictions
- Local Irrigation District/Canal Company - Irrigation water access, water shares, schedules
- County/City Utilities Department - Well permitting, water supply rules

UTILITY & ENERGY PROVIDERS
- Electric Company - Easements, renewable energy programs, line setbacks
- Gas Company - Easements, underground line maps
- Utility Locater Service (e.g. *811) - Underground utility identification before digging

STATE & FEDERAL AGENCIES
- State Department of Agriculture - Cottage food laws, farm stand rules, pesticide regulations
- State Environmental Quality/Natural Resources - Wetlands, stormwater, habitat protection
- USDA / NRCS - Soil Surveys, conservation programs, erosion regulations
- STATE / USGS - Geology, flood maps, hazard maps
- FEMA - Flood map service, floodplain designations/restrictions

PUBLIC HEALTH & SAFETY AGENCIES
- County Health Department - sepctic systems, composting toilets, greywater use
- Local Fire Department - Burn permits, defensible space, fire-safe landscaping

SPECIALTY RESOURCES (DEPENDING ON PROPERTY USE)
- Farmer's Market Websites - Selling homegrown/processed foods requirements
- State Wildlife/Fish & Game - Hunting zones, pond fish regulations, endangered species, wildlife corridors
- Land Conservation Districts - Erosion control, watershed protection rules

SECTION 1:
FUNDAMENTAL DESIGN ASPECTS

Every regenerative & sustainable design begins with creating a base map, setting some goals and boundaries, analyzing the legalities of the land and water, and assessing for assets, risks, and hazards.

Discovering these things at the start not only ensures legal compliance but also saves valuable time and money in the long run.

LEGAL CONCEPTS:

PROFESSIONAL TITLES

LANDSCAPE ARCHITECT: A licensed professional with a degree in landscape architecture.

LANDSCAPE DESIGNER: A professional with training in horticulture and planning who may not be a licensed architect.

LANDSCAPE CONTRACTOR: A professional who performs the installation based on a landscape plan.

PROPERTY RIGHTS AND REGULATIONS

EASEMENT: A legal right to use a part of someone else's land for a specific purpose, such as public utilities.

SETBACK: A required distance from a property line or street where structures, trees, or other landscape features cannot be placed. Setbacks will be shown on property plat maps and construction plans.

RIGHT OF WAY: While related to easements, a right of way is a specific type of easement that grants public access to a path or road across private land.

ZONING: Local government rules that dictate what can be built or done on a piece of property, which often includes specific requirements for landscaping. Common Zone types are, Residential, Commercial, Industrial, Rural, Agriculture, Mixed-Use, & Historic.

REAL PROPERTY: The land itself, along with any permanent structures or anything growing on it, such as trees and landscaping.

CONTRACTS AND AGREEMENTS

SERVICE LEVEL AGREEMENT (SLA): A formal contract between a client and a service provider that details the exact work to be performed, standards, timelines, and responsibilities.

LANDSCAPE DESIGN CONTRACT: A contract specifically for the design phase, which may include scope, cost, and deliverables like design plans and 3D renderings.

CONSTRUCTION CONTRACT: An agreement for the actual building and installation of the landscape, outlining the work to be done, costs, and terms.

PLANNING AND PERMITS

BASEMAP: A to-scale, 2D aerial drawing of a site that acts as the foundation for a design, showing its existing permanent features like property lines, buildings, and water sources. It is a crucial first step in the design process used to record observations and act as a blank canvas for layering more detailed designs as the design progresses.

PLAT MAP: A detailed, scaled map showing the divisions of a piece of land, including boundaries, lot lines, and dimensions. It also indicates streets, easements, public utilities, and sometimes flood zones or other features. These maps are created by licensed surveyors and are often required for real estate transactions, land division, and zoning changes, and are public records.

⟶ ***HOW TO FIND PLAT MAPS:** You can find plat maps online through your county's GIS or assessor's website. For physical or official copies, visit the county recorder's or assessor's office, contact a title company, or hire a land surveyor.

SITE PLAN: A broader plan for the entire development site, of which the landscape plan is a component.

LANDSCAPE PLAN: A formal drawing and document that details the proposed landscape design, often required for obtaining permits.

PERMIT: An official approval from a local government required before starting a project, ensuring that the plans comply with local codes and regulations.

DUE DILIGENCE REPORT: A comprehensive analysis performed before a project begins, which can include evaluating zoning, environmental conditions, and legal encumbrances like easements.

BOUNDARY SURVEY: A survey that accurately determines the legal boundary lines of a property, performed by a professional land surveyor.

REGULATORY SECTORS

Regulatory sectors of government influence landscape design through setting standards for sustainability, managing water and land use, prohibiting harmful chemicals, and establishing rules for public and private spaces. These regulations address ecological concerns like protecting habitats and preventing the spread of invasive species, while also addressing issues like water conservation, public safety, and accessibility. The most common regulatory sectors in relation to landscape design are:

HOME OWNER'S ASSOCIATIONS (HOA's): An organization that makes and enforces rules and guidelines for a residential community. These rules can be extremely strict so always check if the property is tied to an HOA and if it is, read their ordinances and rules.

CC&R'S: CC&R stands for Covenants, Conditions, and Restrictions. These are a set of legally binding rules and regulations, often found in planned communities, that govern the use of a property and its residents' obligations. They are recorded with the local government and typically enforced by a homeowners association (HOA) or the original developer. All HOA's have CC&R's but not all planned communities with CC&R's have HOA's so always check for both.

CITY ORDINANCES: City ordinances are local laws passed by a city's government, such as a city council, to regulate issues that affect the community, like zoning, noise levels, and public safety.

WATER CONSERVATION DISTRICTS: Many areas especially in the desert have city or county water conservation laws that regulate water use. Many also encourage low water landscape designs and remodels through financial compensation incentive programs. To find water conservation laws for your area check your local city or county government website, your local water utility's website, or contact their planning and water conservation departments to see if they have any incentive programs.

STATE & FEDERAL DEPARTMENT OF AGRICULTURE: State and federal agriculture departments regulate a wide range of activities, including food safety and inspection, animal and plant health, agricultural trade, and natural resource conservation. Federal agencies like the USDA handle national and international issues such as setting organic standards, while state agencies focus on implementing and enforcing laws within their borders, especially those related to food and animal health.

IT IS IMPERATIVE TO ROOT EVERY DESIGN IN ALIGNMENT WITH YOUR LOCAL LAWS, REGULATIONS, AND ORDINANCES

CREATING A BASE MAP

A base map is the foundation of a design. It showcases existing features before any modifications and serves as a scribble board for land design, water management, and land data analysis. The base map needs to be to-scale and indicate permanent features that you will need to design around. You will use this base map as a foundation for creativity so make copies as needed as the design refines itself.

1 Get a plat map of the property from the county recorder's website or by finding the property on Google Earth and drawing property boundaries where applicable. Alternatively, you can also hand sketch a drawing using accurate dimensions. Graph paper is helpful for this.

2 Mark directional North in the top or bottom corner. Use another arrow to show the direction of the general slope of the property and label it "Slope". Even properties with a lot of variable slopes and valleys have an overall general slope where all the water drains off the property.

3 Mark the water entrance points. This is anywhere surface water enters the property such as from a neighbor whose property is higher than yours or a wash that runs through the property.
Mark the water exit points. This is where excess water drains off the property. If no exit point exists and flooding is an issue you'll need to create one within the design as it unfolds.

4 Mark:

Fixed trees & shrubs	Gas drops & gas lines
Fixed structures	Existing sprinklers & water systems
Doors & windows of fixed structures	Fixed Parking & pathways
Hose spigots	Rain gutter release points
Wells & other water storage	Overhead power lines
Irrigation channels	Fixed fencing

5 Review legalities and check for zoning restrictions, HOA & CC&R Rules, utility placements such as power poles, electrical boxes, plumbing equipment, septic zones, and easements.

BASE MAP TIPS:

- It is not meant to be perfect nor does it need to look professional. The base map serves as a scribble board so you can draw water flow maps, sector maps, and have an accurate property map to design new pathways and grow zones.

- It only depicts things on the property that are permanent. For example, do not add the shed if the shed is going to be removed to make room for something else.

- If you've got slope or 1+ acres of land you should make sure your base map includes topographical data. You can use topographical maps or images from google earth, but you will want the base map to show the topographical land variations so you can read your land's water flow and microclimates. If you live in an urban or suburban area on less than an acre that appears to have little to no slope it is less important that your base map include this information, but you will still need to visit Google Earth to consider topographical data that could be influencing your property.

- There may be more than 1 water entrance and exit points. Evaluate rainwater run off, neighboring water drainage, irrigation water entrance points, and any slopes or drainages that are channeling the water off the property. You will address the current waterflow and design improvements throughout the design process which may change the water entrance/exit points but to start with you need to identifiy where they are currently.

- Marking doors and windows is important so you don't end up designing something that obstructs views or gets in the way of foot traffic.

BASE MAP EXAMPLE

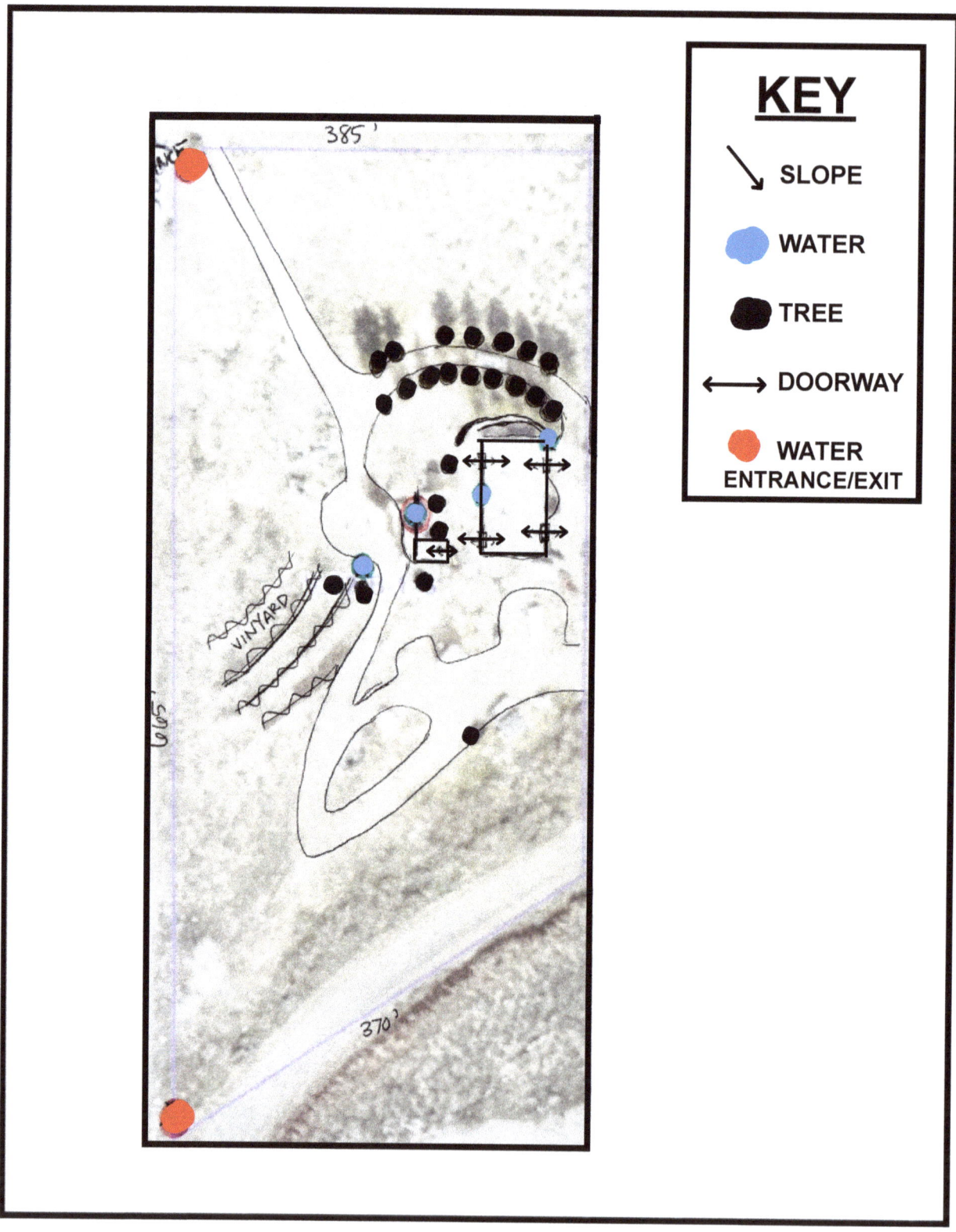

KEY

↘ SLOPE

🔵 WATER

⬤ TREE

↔ DOORWAY

🔴 WATER ENTRANCE/EXIT

385'

370'

VINYARD

PROPERTY QUESTIONNAIRE

Defining your goals and preferences before beginning a landscape design is essential. By clarifying your needs, values, and vision for the land, you ensure that every element of the design has purpose and direction. Clear goals serve as a guiding compass, helping you decide what to plant, where to invest, and how to manage the space over time. Use the following worksheets to outline your goals—this will serve as the master plan of action as your design develops and is put into place.

PROPERTY GOALS

Write down your primary goal for your property. This is usually a simple statement or paragraph describing your overall vision for the property. (for example: Create a low water native plant based landscape, or Design a food forest that feeds our family.

BUDGET:

TIMELINE OR DEADLINE:

FUNCTIONAL NEEDS

List some functions the property design needs to have incorporated into it

- ☐ SHED/ STORAGE
- ☐ BARN/COOP
- ☐ ACCESSIBLE PATHWAYS
- ☐ GREENHOUSE
- ☐ GARDEN BEDS

- ☐ OUTDOOR DINING/ LIVING
- ☐ WATER STORAGE
- ☐ MORE PARKING
- ☐ PLAY AREA
- ☐ MORE PRIVACY

- ☐ _____
- ☐ _____
- ☐ _____
- ☐ _____
- ☐ _____

HOW MUCH TIME ARE YOU WILLING TO SPEND MAINTAINING THE LANDSCAPE?

Annual vegetables are high maintenance plants in comparison to their perennial counterparts. If you do not intend to garden on a daily or weekly basis, limit your installment of annual vegetables and other high maintenance plants. Circle one.

DAILY   **WEEKLY**   **MONTHLY**   **SEASONALLY**

ALLERGENS
To ensure you have a landscape that doesn't create or exacerbate allergies, please list any plant, food, animal, or bee allergens here

PESTS
Regenerative landscapes use plants & other natural techniques to mitigate pests. In order to design pest mitigation into the landscape, list any and all pests that are invasive on the property. (cockroaches, ants, squirrels, racoons, etc)

PRIVACY CONCERNS
Increased privacy can be designed into the landscape. List any areas of concern regarding privacy (example: east fenceline, north bedroom window, etc)

COLOR PREFERENCES
Many plant species have varieties with different colors. If you have a color scheme preference list it here

COOL TONES

WARM TONES

COLOR SCHEME PREFERENCE:_____

PARKING RESTRICTIONS FOR INSTALLERS
If there are neighborhood or property parking restrictions please list them here.
Please specify: ex. no parking Thursdays or 2 hour parking or must park in driveway

HOA or CC&R RESTRICTIONS
List any known HOA or CC&R restrictions pertaining to landscape installation, plant types, and/or maintenance requirements here

HOA NAME: _____

HOA PRESIDENT: _____HOA PHONE NUMBER_____

WEBSITE LISTING HOA RULES/CC&R RESTRICTIONS:_____

RULES:

OTHER CONCERNS
Are there any other specific concerns you would like addressed within the design? (e.g., strong winds, pedestrian traffic, drainage issues, sun exposure, safety, slope issues, etc.)

EDIBLE PLANT PREFERENCES

NON-EDIBLE PLANT PREFERENCES

LIST ANY PLANTS YOU *DON'T* WANT

LEGALITIES

OWNER INFORMATION

NAME:	PHONE:

EMAIL:

PROPERTY ADDRESS:

LEGAL ADDRESS(e.g. K-39-4-51, KIO ESTATES LOT 4):

PROPERTY SIZE:

ZONING CLASSIFICATION

- [ ] RESIDENTIAL
- [ ] AGRICULTURAL
- [ ] RURAL
- [ ] COMMERCIAL
- [ ] INDUSTRIAL
- [ ] HISTORIC
- [ ] MIXED-USE
- [ ] OTHER:

_____

UTILITY COMPANIES

ELECTRICITY:

WEBSITE:	PHONE:

GAS

WEBSITE:	PHONE:

PROPANE:

WEBSITE:	PHONE:

TRASH COLLECTION:

WEBSITE:	PHONE:

WATER:

WEBSITE:	PHONE:

SOLAR PANELS:

WEBSITE:	PHONE:

LEGAL RESTRICTIONS ON USE

These are city codes & ordinances, community regulations, and neighborhood rules such as HOA's & CC&R's that regulate property use, maintenance, & construction. Research your city's codes & ordinances website, HOA documents, & any other relevant community regulations to discover any restrictions or requirements relevant to the design goals.

OUTBUILDING RESTRICTIONS	
FENCING REGULATIONS	
WATER REGULATIONS	
AESTHETIC REQUIREMENTS	
MAINTENANCE & RENOVATION RULES	

ANIMAL RESTRICTIONS PER CITY & NEIGHBORHOOD CODES

SPECIES	QUANTITY ALLOWED	NOTES/EXCEPTIONS

LOCAL COTTAGE FOOD LAWS RELEVENT TO GOALS

PERMITS	
PERMIT TYPE	**INFORMATION**
WELL PERMIT Required if digging a well	FEE:$_____ WEBSITE: _____
STORMWATER MANAGEMENT PERMIT Required for construction projects that disturb 1 acre or more. It controls the discharge of stormwater runoff to prevent erosion and damage	FEE:$_____ WEBSITE: _____
GRADING PERMIT This is often required for land development that significantly changes the property's topography	FEE:$_____ WEBSITE: _____
PLUMBING PERMIT Required for projects involving the installation, modification, or repair of a buildings' water supply or drainage system.	FEE:$_____ WEBSITE: _____
RAINWATER HARVESTING This is specific to systems that collect, store, and distribute rainwater. It addresses system design, installation, and public health standards.	FEE:$_____ WEBSITE: _____
BUILDING PERMIT Required for any new construction	FEE:$_____ WEBSITE: _____
ZONING AND LAND USE PERMIT Occasionally required to ensure proposed water systems comply with the designated zoning regulation for the property	FEE:$_____ WEBSITE: _____
WATER USE Sometimes required when accessing a new or alternative water source, such as a well or rainwater harvesting system.	FEE:$_____ WEBSITE: _____
OTHER:	FEE:$_____ WEBSITE: _____

Be sure to reach out to your city or local professionals if you have any questions about permits. Avoiding the permit vetting process can be costly, time consuming, & can even shut a project down.

WATER LEGALITIES

GREATER WATERSHED:
(This is the major river drainage that leads to the ocean. There are 22 in the United States, 25 in Canada, & 12 in Mexico.)

_____

LOCAL WATERSHED:
(This is the minor tributary that drains into the major river system. They are extremely localized. You can find your local watershed by googling your state or regional watershed maps.)

_____

WATER DISTRICT:
(This is the provider of your water. Check your water bill or use an online search tool to find the company name)

_____

WEBSITE:_____ PHONE:_____

WATER RIGHTS SPECIFICATIONS

TYPE	AMOUNT + ANY SPECIFICATIONS
WATER RIGHTS (ACRE FEET)	
WATER SHARE (IRRIGATION RIGHTS)	
MUNICIPAL (CITY PROVIDED)	
GROUNDWATER/WELL	
LITORRAL	
APPROPRIATIVE	
HYBRID/OTHER	

WATER SHARE / IRRIGATION SPECIFICATIONS (IF APPLICABLE)

IRRIGATION WATER/CANAL COMPANY: (Who manages the system?)

_____WEBSITE:_____

SEASON START DATE:	SEASON END DATE:
IRRIGATION DAY:	IRRIGATION HOURS:
INTAKE LOCATION:	EXIT LOCATION:

WATER QUALITY REPORT

Water quality reports are an important reference for everyone, especially for well water users. If you get your water from a municipal source (city, county, or water district), you can obtain a Consumer Confidence Report (CCR), which provides details on water quality, contaminants, and treatment processes.

WAYS TO GET YOUR WATER QUALITY REPORT:

- Check Your Water Bill or Provider's Website: Most water utilities publish the annual CCR report online (usually by July 1st). The website may have a "Water Quality" or "Consumer Confidence Report" section.

- Use the EPA's Online Tool: Visit EPA's Water System Search and enter your city, county, or water district name to access reports.

- Call Your Local Water Provider: The contact information is on your water bill or the provider's website. Ask for the latest water quality report or request a mailed copy.

- Check with Your County Health Department: Some counties monitor local wells and municipal water and may have additional testing results.

- F*or Well Water Users: Private wells are not regulated, so you'll need to test your own water. Contact your county health department or a certified lab for testing services*.

HOW TO READ A WATER QUALITY REPORT

Physico-chemical indicators are the most common ways to measure water quality. They describe the physical and chemical conditions of water and help us know if it is safe, healthy, and balanced for people, plants, animals, and the environment. Here's what each of the main indicators mean:

INDICATOR	WHAT IT MEANS	WHAT TO LOOK FOR
Dissolved Oxygen (DO):	This shows how much oxygen is available in the water for fish and other aquatic life.	Higher levels are better. Low oxygen means the water may be polluted or stagnant.
PH	measures how acidic or basic the water is. A pH of 7 is neutral; lower numbers are acidic, higher numbers are basic.	Most freshwater life thrives between pH 6.5–8.5. Outside this range, water can stress or harm aquatic species.
TEMPERATURE	Water temperature affects oxygen levels and the health of aquatic life. Warmer water holds less oxygen and may stress fish.	Sudden temperature changes or unusually high/low readings can be a red flag.
SALINITY	Salinity is the amount of dissolved salts in the water. It affects what plants and animals can live there.	Freshwater should have low salinity; higher readings may show contamination or saltwater intrusion.
NUTRIENTS (NITROGEN & PHOSPHEROUS)	These are natural nutrients, but too much can cause algae blooms, which reduce oxygen and harm water life.	Moderate levels are normal, but high levels suggest fertilizer runoff or sewage pollution
TOXICANTS (INSECTICIDES, HERBICIDES, METALS)	These are pollutants that can come from farms, industry, or urban runoff. They may be harmful even in small amounts.	Ideally, reports should show "non-detectable" or very low levels of these substances.

HAZARD ASSESSMENT

Now that you've gathered some foundational design aspects discovered and recorded, it's time to do a hazard assessment. A hazard assessment is an important step in understanding both the strengths and vulnerabilities of a property. Identifying potential risks such as flooding, wildfires, landslides, or seismic activity, allows you to design with greater resilience and safety in mind.

Most states, counties, and regions provide public interactive GIS (Geographic Information System) maps that allow you to explore these hazards in detail. Simply search online for your state or county name along with 'hazard GIS map' or 'interactive hazard map' to find the interactive map for your area. These tools make it possible to view risks specific to your location and integrate that information into your property design.

UTAH'S GEOLOGICAL HAZARDS PORTAL GIS MAP SCREENSHOT

Common hazards that are mapped include:

ROCKFALL ZONES	LANDSLIDE ZONES	HIGH WIND ZONES
EARTHQUAKE FAULT LINES	SHALLOW BEDROCK	TSUNAMI ZONES
LIQUIFACTION ZONES	SHALLOW GROUNDWATER	SINKHOLE ZONES
EROSION ZONES	COLLAPSIBLE SOIL ZONES	FLOODING ZONES

ALWAYS CHECK YOUR LOCAL HAZARD MAPS TO ASSESS WHAT HAZARDS ARE PRESENT SO YOU CAN DESIGN WITH SAFETY IN MIND!

HOW TO UTILIZE YOUR LOCAL HAZARD GIS MAPS

1. **SEARCH FOR THE INTERACTIVE MAP**
In a web browser, type your state or county name + "hazard GIS map" or "interactive hazard map."
Example: "Utah hazard GIS map" or "Clark County interactive hazard map."

2. **OPEN THE INTERACTIVE MAP**
Look for an official source, such as a state emergency management office, county planning department, or natural resources agency.
These sites usually provide a public web map you can click through.

3. **ENTER YOUR PROPERTY ADDRESS**
Use the search bar (address, parcel number, or GPS coordinates) to zoom in on your property.
You can also navigate manually by zooming into the map.

4. **TURN ON HAZARD LAYERS**
Most maps let you toggle different "layers" (flood zones, wildfire risk, fault lines, landslides, etc.).
Check multiple layers to get a complete picture of potential risks.

5. **RECORD YOUR FINDINGS**
Note which hazards apply to your property (floodplain? wildfire-prone area? steep slope?).
Save or screenshot the maps for your records.

With your localized hazards identified, you can now make design decisions that prevent or mitigate hazards while protecting your property and assets.

LOCAL HAZARD ASSESSMENT

Utilize your online county hazard interactive maps to assess for the following hazards. Attach copies of the findings to the property file if desired.

DRAW PROPERTY OUTLINE & NOTE THE AREA OR DIRECTION THE HAZARD IS IN RELATION TO THE LAND.

- ☐ ROCKFALL
- ☐ EARTHQUAKE FAULT LINES
- ☐ LIQUIFACTION
- ☐ FLOODING
- ☐ FIRE
- ☐ EROSION
- ☐ LANDSLIDE
- ☐ WINDBLOWN SAND
- ☐ SHALLOW BEDROCK
- ☐ SHALLOW GROUNDWATER
- ☐ COLLAPSIBLE SOIL
- ☐ EXPANSIVE SOIL
- ☐ TORNADO
- ☐ HURRICANE
- ☐ TSUNAMI

HAZARD NOTES:

_____

_____

_____

_____

PESTICIDES & CHEMICAL EXPOSURE

KNOWN PESTICIDES USED
☐ HERBICIDES
☐ INSECTICIDES
☐ RODENTICIDES

DOES THE CITY SPRAY PESTICIDES ANYWHERE NEAR THE PROPERTY?

YES / NO

IF SO, IS THERE A WAY TO OPT OUT?

YES / NO

DRAW PROPERTY OUTLINE & NOTE DIRECTIONAL WIND AND/OR SLOPES THAT DRAIN ONTO AND EXPOSE THE PROPERTY TO CHEMICALS:

DOCUMENT ALL THAT APPLY

ISSUE	LOCATION
EROSION	
STEEP SLOPE	
POOR SOIL	
FLOODING	
HIGH WINDS	
EXCESSIVE SUN/HEAT TRAPS	
EXCESSIVE SHADE / NO SUN	
LACK OF PRIVACY	
WILDLIFE INTRUSIONS	
TOXIC CHEMICAL EXPOSURE ZONES	
EXCESSIVE NOISE	
OTHER	

THREAT & ASSET ANALYSIS

Every property has threats and assets. Evaluating this allows for the design to include threat mitigation and ideas that enhance the valuable or profitable assets on site. Assess your property and check the boxes in the following tables that apply to your land.

COMMON NATURAL ASSETS

✓	ASSET TYPE	EXAMPLES	NOTES
	WILDLIFE HABITAT	Birds, pollinators, frogs, lizards, beneficial snakes, bats	Supports pest control, pollination, nutrient cycling.
	ESTABLISHED TREES	Fruit trees, nut trees, mature shade trees, nitrogen fixers (e.g., mesquite, black locust)	Offer shade, microclimate buffering, carbon sequestration, soil improvement.
	HEALTHY SOIL	Loamy texture, good structure, active biology, fungal presence	Increases water-holding, fertility, and root health.
	TOPSOIL DEPTH	Thick topsoil, low compaction, visible organic matter	Saves years of soil-building work.
	GRAVITY FED WATER SOURCE	Spring, seep, uphill water tank, or pond	Enables passive irrigation without pumps.
	NATURAL POND OR WETLAND	Supports aquaculture, wildlife, filtration, and microclimate stability	Can become the heart of a regenerative design.
	GOOD DRAINAGE	Slight slope, no waterlogging	Reduces erosion and root disease.
	SOUTHERN OR EASTERN EXPOSURE	Ideal slope/aspect for sun-loving crops	Extends growing season and increases solar energy capture.

	COOL MICROCLIMATES	North-facing slopes, shaded zones, canyon walls	Allows for planting of more delicate crops in hot zones.
	SAFE NEIGHBORHOOD	Places with high safety records	Increases property values and property safety overall
	FROST FREE ZONES	Thermal mass near boulders, ponds, walls, and full sun locations	Extends the growing season or allows for subtropical plants in marginal climates.
	ONSITE MULCH SOURCES	Pine needles, leaves, wood chips, straw, invasive biomass	Reduces cost of sheet mulching and improves soil fertility.
	LOCAL CLAY OR STONE	For cob, earthworks, or dry-stacking	Valuable for natural building or erosion control.
	EXISTING FENCING	Perimeter or cross-fencing	Reduces infrastructure costs for animals, gardens, or zones.
	EDGE ACCESS TO WILDLANDS	Forest, desert, BLM land, wildlife corridor	Increases biodiversity and foraging potential.
	BENEFICIAL NEIGHBORS	Apiaries, seed savers, fellow permaculturists, regenerative farmers	Allows for bartering, learning, and community support.
	LOCAL RAINFALL CAPTURE POTENTIAL	Rooflines, roadways, or slopes suitable for water harvesting	Enables water self-sufficiency and erosion repair.
	ABUNDANT NATURAL MATERIALS	Downed wood, rock piles, woody weeds, bamboo, reeds	For hugelkultur, trellising, mulch, shelter building, etc.

	NATIVE PLANTS OR WILD MEDICINE	Native herbs, berries, fruit, nuts, etc.	Adds ecological value, food, medicine, and restoration potential.
	GOOD ACCESS ROADS	Drivable in all weather, no washouts or erosion	Reduces energy costs and makes implementation safer and easier.
	CLEAN AIR & OPEN SKY	Low pollution, good solar access	Promotes plant health and improves solar panel viability.
	GREAT VIEWS	Open vistas, cityscapes, or other desirable views	Increase property values & Ambiance
	ESTABLISHED WELL	Well on site	Water security and overall money saver
	ESTABLISHED SOLAR	Solar panels and accompanying equipment are established on the property	Energy security and overall money saver
	OTHER		
	OTHER		
	OTHER		

COMMON NATURAL THREATS

✓	ASSET TYPE	EXAMPLES	NOTES
	WILDLIFE CONFLICTS	Raccoons, deer, rabbits, gophers, voles, wild boars, coyotes, snakes, skunks, porcupines	May eat crops, damage structures, or pose safety risks. Consider fencing or habitat buffers.
	PREDATORY BIRDS	Hawks, owls, eagles (if raising poultry or small animals)	Can be a threat to chickens, ducks, or small pets.
	INSECTS & ARACHNIDS	Fire ants, wasps, mosquitoes, termites, ticks, scorpions	Affect humans, pets, structures, or livestock.
	POISONOUS PLANTS	Poison ivy/oak/sumac, oleander, hemlock, jimsonweed, castor bean	Risk to humans, pets, and livestock
	WIND	Seasonal high winds, tornadoes, microbursts	Can damage trees, structures, or desiccate soil.
	FIRE RISK	Dry brush, dead trees, surrounding forest, prevailing wind corridors	Wildfire zones or unmanaged fuels. Firewise strategies may be needed.
	FLOODING	Seasonal stream overflow, pond breaches, water table fluctuations, flash floods	Especially a concern near washes or in clay-heavy soils.
	EROSION	Bare slopes, gullies, compaction runoff	Undermines soil health and long-term sustainability.
	LANDSLIDES	Steep slopes with unstable subsoil or poor drainage	Relevant in hilly or mountainous areas.

	SOIL CONTAMINATION	Heavy metals, pesticide residue, salt buildup, petroleum contamination	Can result from past land use, nearby roads, or mining.
	DROUGHT CONDITIONS	Long-term water shortages, low aquifer recharge, municipal water restrictions	Threatens water resilience.
	WATER QUALITY ISSUES	High salinity, alkalinity, sulfur, nitrates, or biological contamination	Especially critical if relying on wells, springs, or ponds.
	INVASIVE SPECIES	Bindweed, goathead, cheatgrass, Johnson grass, tamarisk, Russian olive, etc.	Outcompete natives and require consistent management.
	WEED PRESSURE	Windblown seed (e.g., thistle, mustard)	High edge exposure = higher weed pressure.
	MICROCLIMATE EXTREMES	Frost pockets, reflected heat zones, hot slopes, excessive shade	May limit growing options or create unexpected stress zones.
	POLLUTION DRIFT	Pesticide or herbicide drift from neighboring farms or roads	Risk to organic practices, pollinators, or sensitive plants.
	NEIGHBORING LAND USE	Feedlots, shooting ranges, dog kennels, negligent landowners	May impact water, sound, air quality, or wildlife habitat.
	BEE ALLERGIES	Nearby apiaries or wild hives	Can be a health risk for allergic individuals, & large apiaries can be damaging to local native bee populations

BIGGEST THREATS TO DESIGN SOLUTIONS FOR
(Don't worry about how for now, just list the threats relevant to your property.)

BIGGEST ASSETS TO NURTURE, AMPLIFY, OR PROFIT OFF OF

GATHERING RESOURCES!

Using local materials supports the local economy, reduces the environmental impact of transportation, and ensures that the materials are well-adapted to the regional climate and ecology which is especially important to beneficial insects & the pollinator community & cycle at large.

Local resources like native stone, clay, mulch, and plants often require less maintenance and integrate more harmoniously with the land, making your design more resilient and cost-effective.

Additionally, sourcing locally fosters community relationships and reduces dependency on global supply chains. It ensures skill, value, and money stays within the community benefiting everyone at large instead of big corporations.

LOCAL PROFESSIONAL TRADES
(TRY TO FIND INDEPENDENT CONTRACTORS OR SMALL LOCAL BUSINESSES)

LOCAL PLUMBER

NAME:_____
COMPANY:_____
WEBSITE:_____
PHONE NUMBER:_____

LOCAL ELECTRICIAN

NAME:_____
COMPANY:_____
WEBSITE:_____
PHONE NUMBER:_____

LOCAL MECHANIC

NAME:_____
COMPANY:_____
WEBSITE:_____
PHONE NUMBER:_____

LOCAL WELDER

NAME:_____
COMPANY:_____
WEBSITE:_____
PHONE NUMBER:_____

LOCAL CARPENTER

NAME:_____
COMPANY:_____
WEBSITE:_____
PHONE NUMBER:_____

LOCAL LANDSCAPER

NAME:_____
COMPANY:_____
WEBSITE:_____
PHONE NUMBER:_____

LOCAL HANDYPERSON

NAME:_____
COMPANY:_____
WEBSITE:_____
PHONE NUMBER:_____

LOCAL HVAC

NAME:_____
COMPANY:_____
WEBSITE:_____
PHONE NUMBER:_____

LOCAL MATERIAL RESOURCES

LOCAL NURSERY
COMPANY:_____
WEBSITE:_____
PHONE NUMBER:_____

LOCAL NURSERY
COMPANY:_____
WEBSITE:_____
PHONE NUMBER:_____

ETHICAL SEED COMPANIES *Try to source seeds from your regional watershed*

COMPANY:_____
WEBSITE:_____
SOCIAL MEDIA:_____

COMPANY:_____
WEBSITE:_____
SOCIAL MEDIA:_____

COMPANY:_____
WEBSITE:_____
SOCIAL MEDIA:_____

COMPANY:_____
WEBSITE:_____
SOCIAL MEDIA:_____

LOCAL LANDFILL
COMPANY:_____
LOCATION:_____
WEBSITE:_____
PHONE NUMBER:_____

LOCAL COMPOST
COMPANY:_____
LOCATION:_____
WEBSITE:_____
PHONE NUMBER:_____

LOCAL MANURE
COMPANY:_____
LOCATION:_____
WEBSITE:_____
PHONE NUMBER:_____

LOCAL WOODCHIPS/ARBORIST
COMPANY:_____
LOCATION:_____
WEBSITE:_____
PHONE NUMBER:_____

LOCAL LUMBER YARD
COMPANY:_____
LOCATION:_____
WEBSITE:_____
PHONE NUMBER:_____

LOCAL QUARRY
COMPANY:_____
LOCATION:_____
WEBSITE:_____
PHONE NUMBER:_____

LOCAL LOCAL HARDWARE STORE
COMPANY:_____
LOCATION:_____
WEBSITE:_____
PHONE NUMBER:_____

LOCAL AGRICULTURE STORE
COMPANY:_____
LOCATION:_____
WEBSITE:_____
PHONE NUMBER:_____

SECTION 2:
SYSTEMS THINKING ASSESSMENT

Systems thinking is the practice of understanding how all aspects of your property: climate, topography, biology, geography, and ecology, are connected into one whole system.

Rather than looking at each part in isolation, systems thinking emphasizes relationships, feedback loops, and patterns of interaction. This perspective lies at the heart of permaculture design, where every element is placed with intention to serve multiple purposes and connect with others.

By thinking in systems, you can design landscapes that cycle energy efficiently, minimize waste, and become greater and more resilient than the sum of their parts.

SYSTEMS THINKING CONCEPTS

3 FOOT RULE: An average adult has a 3 foot reach and most wheelbarrow pathways are 3 feet wide. Plant and design accordingly.
RELEVANCE: design grow beds no larger than 6 feet across so plants are easy to reach from all sides. If grow areas are wider than that, stepping stones or pathways will need to be added for internal access.

4 SEASON OBSERVATION: Observing the land over all 4 seasons and through all major weather happenings prior to design or development.
RELEVANCE: Seasonal changes and localized micro weather patterns can have drastic effects on land and plant growth. Being able to observe the land for 4 seasons prior to or during the design phase is a crucial step for long term resiliency & stability.

CLOSED LOOPS: Waste becomes a resource within the system.
RELEVANCE: Encourages composting, water reuse, and zero-waste practices.

FEEDBACK LOOPS: Observing for positive and negative feedback cycles in order to regulate ecosystems (e.g., predator-prey relationships).
RELEVANCE: Helps in understanding dynamic systems and designing resilient ecosystems. For example, noticing a pest infestation and working to encourage it's natural predator to combat the issue.

SECTOR MAP: A sector map is a permaculture planning tool used to analyze external influences on a landscape, such as sun, wind, wildlife, hazard exposure, and human activity.
RELEVANCE: It helps by highlighting which forces are present & where they are coming from thereby helping you to mindfully design resilient, efficient, and sustainable systems by working with natural patterns instead of against them.

ZONES OF USE: A design system that allows you to place items and design systems with respect to how often it needs to be tended.
RELEVANCE: Placing high maintenance daily/weekly visited gardens close to the home increases visibility and keeps pests away via all the human traffic.

THE SCALE OF PERMANENCE- A design tool that ranks landscape elements based on how difficult they are to change, from the most permanent to the most flexible.

RELEVANCE: *Originally developed by P.A. Yeomans, By designing from permanent to flexible you ensure that more permanent features like water flow and roads are placed wisely before designing where the trees or structures go that are affected by them.*

CLIMATE   LANDFORM   WATER   ACCESS   VEGETATION   STRUCTURES   SOIL

MOST PERMANENT   **MOST FLEXIBLE**

THE IMPORTANCE OF THE SCALE OF PERMANENCE:

① It Prevents Costly Mistakes

Making design decisions out of order can lead to expensive and time-consuming errors.The Scale of Permanence ensures you establish the foundational systems such as water, access, and structures before the living systems that depend on them.

For example: You wouldn't plant your orchard before managing drainage or access.

② It Creates Resilience and Efficiency

When each design layer is built upon stable foundations, the resulting system is far more resilient. By designing from permanent to temporary, you create landscapes that work with natural processes instead of fighting them, leading to long-term stability, lower maintenance, and greater ecological harmony.

③ It Encourages Holistic Thinking

The Scale of Permanence is not just a checklist. It's a thinking tool. It helps designers visualize how each element affects the others.

For instance, understanding microclimates (influenced by landform and water) helps determine optimal building placement or plant guild design.

④ It Supports Regenerative, Not Just Sustainable, Design

Because it respects natural patterns of permanence and change, the Scale of Permanence encourages systems that improve over time rather than merely sustain themselves. Designers use it to integrate water harvesting, soil building, and vegetation planning in ways that regenerate the landscape year after year.

SECTOR MAP

A sector map is a permaculture planning tool used to analyze external influences on a landscape, such as sun, wind, wildlife, hazard exposure, and human activity. Sector maps highlight which forces are present & where they are coming from thereby helping you to mindfully design resilient, efficient, and sustainable systems by working with natural patterns instead of against them.

Steps to Creating a Sector Map

 **OBTAIN OR SKETCH YOUR BASEMAP**
Start with your base map. You can also use plat maps or aerial photos.

IDENTIFY NATURAL AND EXTERNAL FORCES
Analyze the following major external environmental influences that impact your land:

- **SUN PATH** - Identify where the sun rises and sets throughout the seasons. Mark the summer solstice sun path & the winter solstice sun path.
- **PREVAILING WINDS** - Mark strong seasonal winds and natural windbreaks. Be sure to mark the winter winds & the summer winds separately.
- **HAZARDS -** Show any hazards and the direction they are coming from. For example: Rock/mud slide, flooding etc.
- **WILDLIFE & PESTS** - Track frequent animal paths, pest-prone areas, and beneficial habitats.
- **FIRE RISK ZONES** - Identify areas prone to wildfires and potential buffer zones.
- **VIEWS & NOISE POLLUTION** - Consider scenic views to preserve and noise sources to mitigate.

Use different colored arrows or shading to illustrate each sector.
Example:
Yellow arrows for sun exposure, blue arrows for prevailing wind directions, red dotted lines for fire risks, & green areas for existing vegetation or wildlife corridors.

The Sector Map can be as detailed or simple as you like. Reference the Sector Map as necessary throughout the design process.

SECTOR MAP EXAMPLES

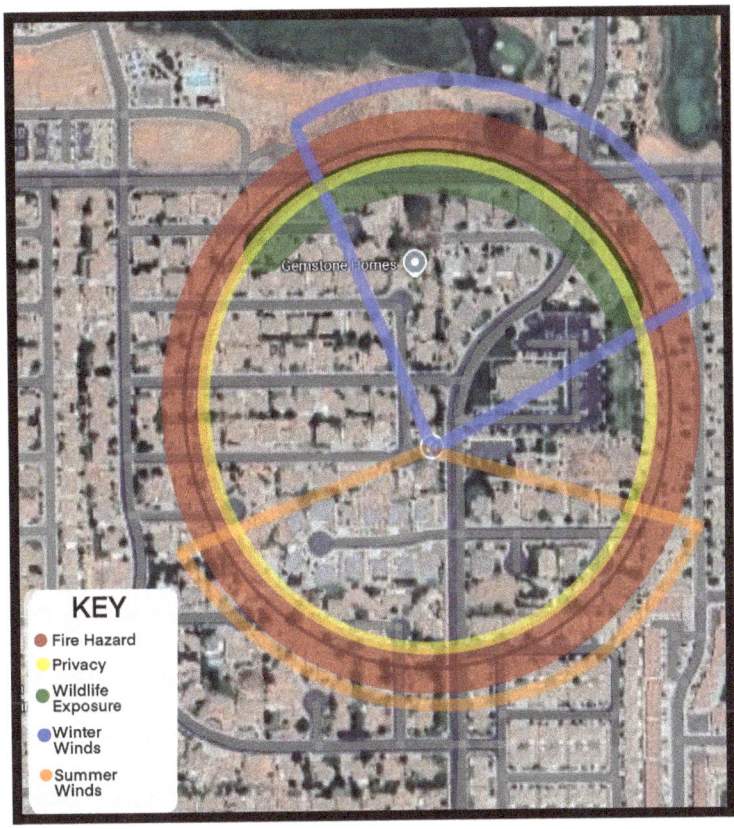

KEY
- Fire Hazard
- Privacy
- Wildlife Exposure
- Winter Winds
- Summer Winds

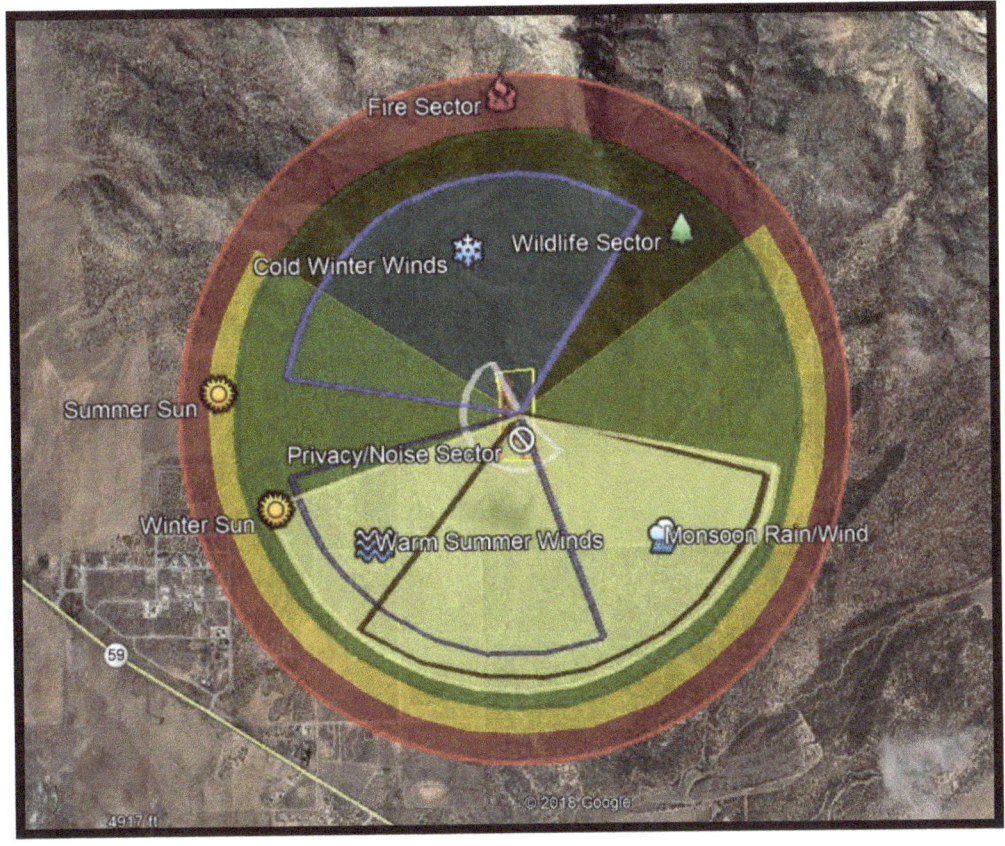

SECTOR MAP

Use your basemap to create a Sector map for your property. It can be as simple or as detailed as you prefer.

ZONES OF USE

The Zones of Use is a permaculture concept that helps to organize a landscape based on frequency of human activity and energy use. The closer a zone is to the home, the more frequently it is used and managed. For example, vegetable and herb gardens are frequented often. Therefore they should be placed closer to the home. Fruit trees and orchards only need seasonal care so they can be placed farther from the home.

PRO TIPS

● Zone 5, the wildlife/native zone, is imperative for pollinators & predator/pest balance. It acts as a safe, chemical free, native cooridor that allows birds, bees, butterflies, & other beneficial wildlife to use as a home or pitstop as they travel from place to place. If you live in an urban area zone 5 may just consist of your fence line, but no matter how much land you can spare for zone 5 you should dedicate that strip to native plants and trees.

● The Zones of use do not need to be linear or even. Some properties may have very little of one zone and a lot of another. The idea is for you to place elements within their zone based on their frequency of use to save time and energy and to create sustainable flow.

ZONES OF USE EXAMPLE CHART

	ZONE 1 Daily	ZONE 2 Weekly	ZONE 3 Monthly	ZONE 4 Seasonally	ZONE 5 Unmanaged
USES	Social space, Play places, Garden beds, Dining space	Garden beds, Berry patches, Play places, Pools	Cash crops, Storage, Beehives, Pasture	Firewood, Lumber, Pasture, Storage	Foraging, Inspiration, Observation
STRUCT-URES	Playground, Greenhouse, Outdoor dining, Chicken Coop, Animal pens	Greenhouse, Compost, Barn, Shed, Animal pens	Feed, storage, Beehives, Fenced pastures	Fenced pastures, Beehives	None
PLANTS	Herbs, Annuals, Vegetables, Lawn	Shrubs, Trees, Berry patches, Perennials	Orchards, Pastures, Native plants	Firewood, Timber, Coppicing trees, Native plants	Native plants
WATER	Rain barrels, Water features, Greywater, Plumbing	Wells, Ponds, Irrigation, Swales	Large ponds, Swales, Irrigation	Ponds, Swales	Lakes, Creeks, Streams, Washes
ANIMALS	Dogs, Cats, Rabbits, Chickens	Small grazing animals	Large grazing animals	Large grazing animals	Wildlife

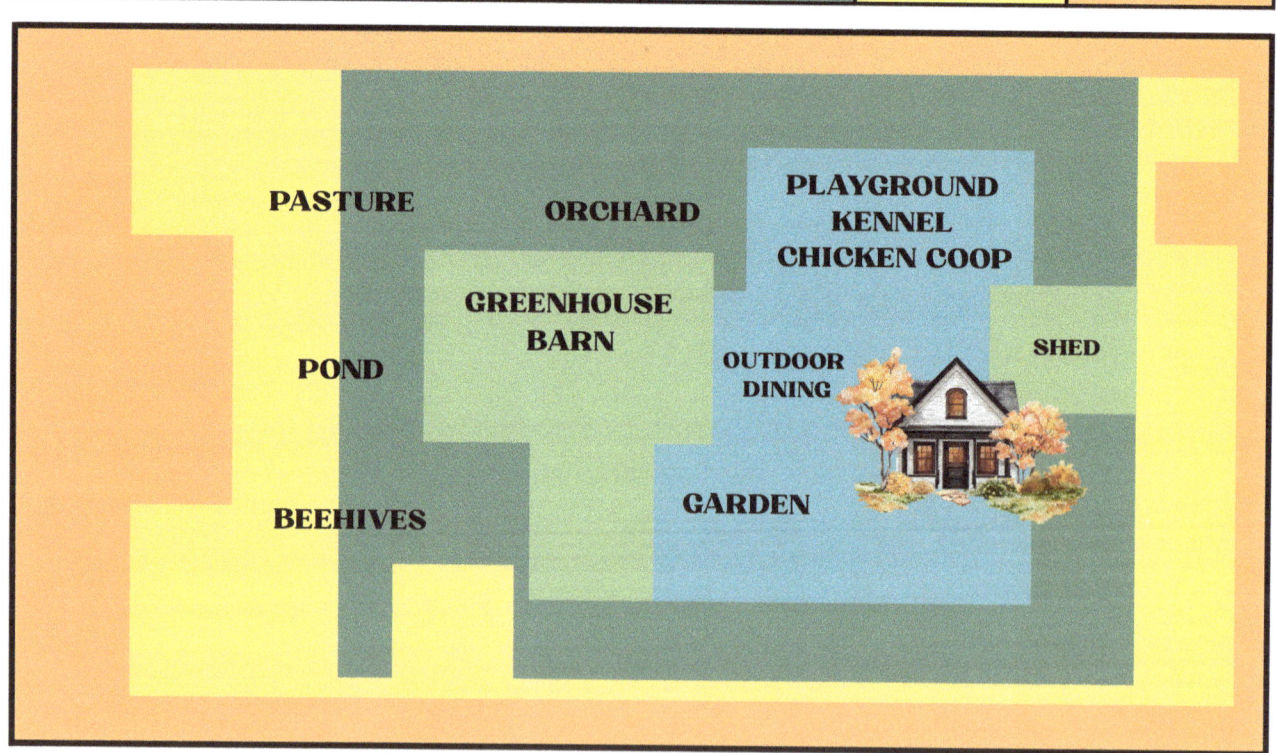

47

ZONES OF USE MAP

Use your basemap to create a Zones of Use map for your property. It can be as simple or as detailed as you prefer. Be sure to place gardens, animal shelters, and water management systems where they make sense for frequent access. If you have a large property, leave buffer areas for windbreaks, wildlife corridors, and future expansion.

N

CLOSED LOOP ASSESSMENT

✓	WASTE OUTPUT	REPURPOSED AS	SYSTEM LOOP EXAMPLE
	COFFEE GROUNDS	Plant fertilizer, mushroom substrate, compost starter	Kitchen → Garden → Compost → Soil health
	KITCHEN VEGETABLE SCRAPS	Worm bin (vermicompost), compost, chicken feed	Kitchen → Worm bin → Garden soil
	EGGSHELLS	Calcium for soil, chicken feed supplement	Kitchen → Crushed → Garden/chicken coop
	WOOD ASH (FROM FIREPLACE/STOVE)	Garden lime, slug repellent, mineral source	Fireplace → Garden/Fruit trees
	GREYWATER (FROM SINKS/SHOWERS)	Irrigation for trees and shrubs (after filtering)	Bathroom → Reed bed → Garden/Orchard
	TOILET COMPOST (FROM COMPOSTING TOILET)	Humanure compost (after 1-2 years aging)	Toilet → Compost → Non-edible landscaping trees
	MANURE (GOATS, CHICKENS, HORSES)	Fertilizer, hot composting, methane (in biodigesters)	Animal shelter → Compost → Garden/Farm
	SPOILED MILK/FRESH MOZZARELLA WHEY	Soil microbe boost, compost additive, animal feed	Kitchen → Garden soil → Increased microbial activity
	YARD WASTE (LEAVES, BRANCHES)	Mulch, hugelkultur, biochar	Yard → Garden bed → Soil structure/water retention
	TREE PRUNINGS	Animal fodder (for goats/rabbits), mulch, wattle fencing	Orchard → Fodder/Fence materials
	OLD WOOL OR NATURAL FABRIC	Weed barrier, nesting material	Closet → Garden/Chicken coop

	USED COOKING WATER (E.G., VEGGIE WATER)	Liquid fertilizer (cool and apply)	Kitchen → Garden
	ANIMAL BONES	Bone broth, then crushed for bone meal	Kitchen → Food → Fertilizer
	SPOILED PRODUCE	Chicken/pig feed, compost, fermentation (bokashi)	Garden/Kitchen → Chickens/Compost
	CHICKEN FEATHERS	Compost, insulation, pillow stuffing	Coop → Garden or home crafts
	CARDBOARD/PAPER WASTE	Sheet mulch, worm bedding, fire starter	Mailbox → Garden → Soil
	RAINWATER FROM ROOFS	Irrigation, livestock water	Roof → Barrel → Garden/Animals
	WEEDS AND INVASIVE PLANTS	Compost, fermentation liquid (weed tea), mulch	Garden → Barrel → Nutrient liquid
	OLD HAY/STRAW BEDDING	Compost, mulch	Animal shelter → Garden
	USED COOKING WATER (E.G., VEGGIE WATER)	Liquid fertilizer (cool and apply)	Kitchen → Garden
	ANIMAL BONES	Bone broth, then crushed for bone meal	Kitchen → Food → Fertilizer
	OLD WOOL OR NATURAL	Fabric, weed barrier, nesting	Closet → Garden/Chicken coop

WHAT GOES WHERE?

NORTH SIDE (Shady, Cool, Protected)
BEST FOR: SHADE-LOVERS, WINDBREAKS, STORAGE, COOL-WEATHER NEEDS
Mushrooms (e.g. wine caps, shiitake on logs)
Compost bins (slower breakdown, but protected)
Water storage tanks (less evaporation)
Firewood storage (keeps dry and shaded)
Animal shelters (protected from harsh summer sun)
Cool-season greens (e.g. lettuce, spinach in summer)
Shade-tolerant herbs (e.g. mint, lemon balm, chives)
Cold storage/root cellar entrance
Beehives (placed facing southeast, but protected from harsh afternoon sun)
Windbreak trees (if prevailing winds are from the north in your region)

WEST SIDE (Hot Afternoon Sun, Intense Light)
BEST FOR: HEAT-TOLERANT, SUN-LOVING PLANTS AND PROTECTIVE FEATURES
Sunflowers, corn, squash, melons (love strong afternoon sun)
Trellised plants as shade screens (e.g. cucumbers, gourds)
Solar dehydrators (strong sun for drying)
Heat-loving herbs (e.g. sage, thyme, lavender)
Barns or sheds (block late-day heat from house or gardens)
Thermal mass features (rock walls, water barrels to absorb heat)
Privacy hedges/windbreaks (slow drying winds, summer sun buffer)

EAST SIDE (Morning Sun, Afternoon Shade)
BEST FOR: DELICATE PLANTS, EARLY WARMTH WITHOUT SCORCHING HEAT
Berries (e.g. raspberries, strawberries, blueberries)
Fruit trees that bloom early (e.g. apricots, cherries—less heat stress)
Children's play area (morning play before it gets too hot)
Medicinal & culinary herbs (e.g. chamomile, calendula, lemon balm)
Pollinator gardens (bees are active earlier in morning sun)
Bees and poultry (morning warmth, afternoon rest)
Garden benches, meditation spots (comfortable morning light)

SOUTH SIDE (Warmest, Full Sun All Day)
BEST FOR: SUN-LOVERS, HIGH-PRODUCTION AREAS, SOLAR GAIN
Orchards & fruit trees (e.g. apples, peaches, figs)
Vegetable gardens (main production area)
Herb spirals or beds (rosemary, basil, oregano, etc.)
Chicken coops/animal pens (winter sun access)
Solar panels or solar water heaters
Greenhouses & cold frames
Drying racks (herbs, clotheslines)
Grapevines & trellises
Washing stations (dry fast in sun)
Open gathering areas for humans (sunny patios, outdoor kitchens)

SECTION 3:
METEOROLOGY ASSESSMENT

Meteorology deals with the climate and weather patterns that shape how your land functions day by day and season by season. It includes factors such as temperature swings, rainfall, wind, humidity, and seasonal extremes.

These elements directly determine which plants will thrive, how water moves across the property, and what risks such as frost, drought, or storms your design must take into account.

By studying and recording local meteorological data, you gain the knowledge to design landscapes that are resilient, efficient, and better prepared for the long term realities of your environment.

METEOROLOGICAL CONCEPTS

ALBEDO EFFECT: The reflectivity of surfaces, affecting how much heat is absorbed or reflected.
RELEVANCE: Used to select surface materials that moderate temperature and create comfortable microclimates. For example, white stone reflects heat, while dark paint & bare soil absorbs it. This can be used to design microclimates, create comfortable patios & pathways, as well as energy efficient structures.

CLOUD COVER PATTERNS: The extent and frequency of cloud presence, which affects light and heat availability.
RELEVANCE: Influences solar energy gain, evaporation rates, and frost risk which is important when designing solar orientation and crop protection strategies.

CONVECTION CURRENTS: Upward air movement caused by heated surfaces, increasing wind and drying potential.
RELEVANCE: Inform placement of vegetation and shaded surfaces to reduce desiccation and erosion.

DIURNAL TEMPERATURE SHIFT: The difference between daytime high and nighttime low temperatures, often extreme in deserts.
RELEVANCE: Guides placement of thermal mass elements like rock walls or adobe to buffer daily temperature swings and support plant health.

DEW POINT & RELATIVE HUMIDITY: Measures of air moisture that impact plant transpiration, disease risk, and comfort.
RELEVANCE: Affects irrigation scheduling, greenhouse humidity control, and disease prevention; especially important in drylands with sharp humidity fluctuations.

MOUNTAIN WINDS / KATABATIC / CANYON WINDS: Cold winds that descend from mountains or flow through canyons, especially at night.
RELEVANCE: Inform placement of windbreaks, crop protection, and structure orientation in high-deserts and mountainous terrain.

THERMAL MASS: Materials that store and slowly release heat (e.g., stone, water).
RELEVANCE: Used in greenhouses, buildings, and ponds for temperature regulation.

NOCTURNAL RADIATION LOSS: Heat escapes quickly at night due to clear skies, causing very cold early mornings.
RELEVANCE: Calls for thermal mass, cold protection for plants, and strategic siting of greenhouses and animal shelters.

DUST STORMS (HABOOBS): Large, fast-moving walls of dust driven by wind in desert environments.
RELEVANCE: Requires the use of ground cover, windbreaks, and living barriers to protect plants and prevent topsoil loss.

FLASH FLOOD RISK: Sudden, high-volume water flow due to impermeable soils and short, intense storms.
RELEVANCE: Informs placement of buildings swales, gabions, ponds, spillways, and planting zones to slow, spread, and safely infiltrate stormwater.

FOG FORMATION: Low-lying clouds formed by condensation, common near coasts or mountains.
RELEVANCE: In coastal deserts or highlands, fog can be harvested or used to support fog-dependent plants like succulents and native shrubs.

FROST POCKETS: Cold air sinks into low spots, leading to higher frost risk in valleys or depressions. Always remember: cold air sinks, hot air rises.
RELEVANCE: Avoid placing sensitive plants or structures in these zones; use thermal mass or topography to buffer cold air.

HEAT ISLAND EFFECT: Environments that trap heat due to concrete, asphalt, roofing, dark surfaces or low vegetation. Common in urban environments.
RELEVANCE: Use trees, reflective surfaces, green roofs, and water features to cool permaculture sites in urban or semi-urban desert settings.

LOW HUMIDITY: A measure of dry air that increases evaporation and stress on plants and soil.
RELEVANCE: Calls for mulching, deep watering techniques, and drought-resilient species to reduce water loss and stress.

MICROCLIMATES: Localized zones with unique weather conditions due to variations in elevation, structures, or vegetation.
RELEVANCE: Leverage these zones to grow diverse plants.

MONSOONS: Seasonal, intense thunderstorms that occur in arid climates like the American Southwest.
RELEVANCE: Crucial for water-harvesting system design, soil retention features, and summer crop planning.

PREVAILING & SEASONAL WINDS: Dominant wind directions that influence evaporation, plant damage, and comfort.
RELEVANCE: *Used to design shelterbelts, windbreaks, orient buildings, and protect garden zones from desiccating wind.*

The three main prevailing winds on the planet are the polar easterlies, westerlies, and trade winds.

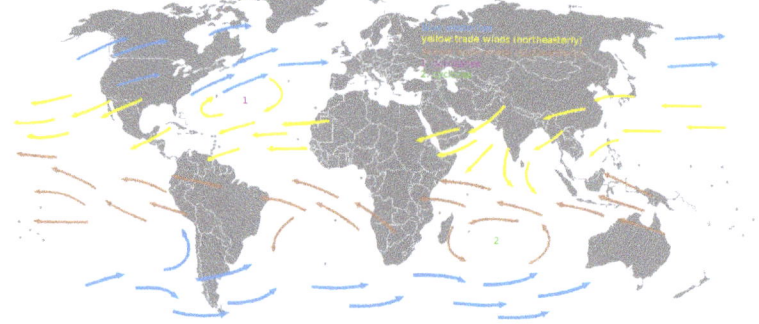

 Trade Winds: These blow from the east toward the west in the tropics, between roughly 0° and 30° latitude in both hemispheres. They move from the subtropical high-pressure zones toward the equatorial low-pressure zone and are steady, warm, and moist.

 Westerlies: These blow from the west toward the east between about 30° and 60° latitude in both hemispheres. They dominate the temperate zones and drive much of the weather in North America, Europe, and southern Australia.

 Polar Easterlies: These cold, dry winds blow from the east toward the west between 60° and 90° latitude in both hemispheres, moving away from the polar high-pressure areas toward the subpolar lows.

RAIN SHADOW EFFECT: Dry conditions created on the leeward side of mountains where moisture is blocked.
RELEVANCE: *Influences long-term planning for irrigation, plant selection, and understanding regional water scarcity. Extremely relevant when choosing property.*

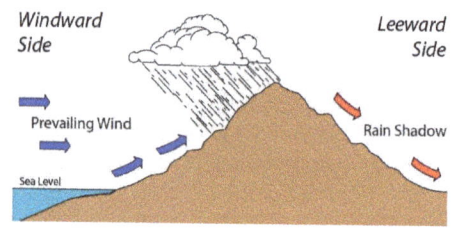

SOLAR PATH & INTENSITY: The seasonal path the sun takes across the sky, influencing light, shade, and heat patterns.
RELEVANCE: *Determines optimal placement for gardens, solar panels, and passive solar buildings.*

RAINWATER BUDGET: A calculation that measures how much rainwater falls on a property and how that water can be captured, stored, and used within the landscape.
RELEVANCE: *It helps you to design water-harvesting earthworks like swales and match plant water needs to what the landscape can naturally supply.*

PROPERTY CLIMATE DATA

Researching local climate data is crucial for creating successful & sustainable landscapes. Ignoring this data can lead to inefficient designs and wasted resources. This data is foundational to plant selection, water management, & analyzing environmental impacts on the land in order to design resiliency & efficiency into the landscape.

USING LOCAL WEATHER DATA, FILL IN THE FOLLOWING DATA FOR YOUR PROPERTY LOCATION

USDA GROW ZONE:	AVERAGE ANNUAL RAINFALL:	AVERAGE ANNUAL SNOWFALL:
_____	_____	_____

AVERAGE MONTHLY RAINFALL

JAN	FEB	MAR	APR	MAY	JUN	JUL	AUG	SEP	OCT	NOV	DEC

+

AVERAGE MONTHLY SNOWFALL

JAN	FEB	MAR	APR	MAY	JUN	JUL	AUG	SEP	OCT	NOV	DEC

=

TOTAL AVERAGE MONTHLY PRECIPITATION

JAN	FEB	MAR	APR	MAY	JUN	JUL	AUG	SEP	OCT	NOV	DEC

AVERAGE MONTHLY HIGHS AND LOWS

HIGH												
LOW												

Diurnal Temperature Range is a measure of how much the temperature varies over a 24-hour period. To get the average monthly diurnal temperature range, subtract the HIGH and LOW temperature data for each month. Input the results below.

DIURNAL TEMPERATURE RANGE												
	JAN	FEB	MAR	APR	MAY	JUN	JUL	AUG	SEP	OCT	NOV	DE
DTR°												

RECORD HIGH		RECORD LOW	
TEMPERATURE:		TEMPERATURE:	
DATE:		DATE:	

AVERAGE HUMIDITY			
SPRING %	SUMMER %	AUTUMN %	WINTER %

SEVERE WEATHER SEASONS		
✓	SEVERE WEATHER	SEASON DURATION IN MONTHS (JUL-OCT)
	MONSOON SEASON	
	HURRICANE/TYPHOON SEASON	
	TORNADO SEASON	
	HIGH UV SEASON	
	SEVERE STORM SEASON (Blizzards, lighting risk, wind, hail, etc.)	
	EXTREME HEAT OR HUMIDITY SEASON	
	OTHER (Mud, fog, avalanche, migrations, etc.)	

PREVAILING WINDS

Draw a directional arrow of the seasonal winds for Winter & Summer. Spring & autumn winds are less predictable, less prevailing, & tend to swing directions during the shift to the more stable winter & summer prevailing winds. Generally speaking, in the northern hemisphere the winter winds come from the north, northwest, or northeast, & the summer winds come from the south, southwest, or southeast depending on your location.

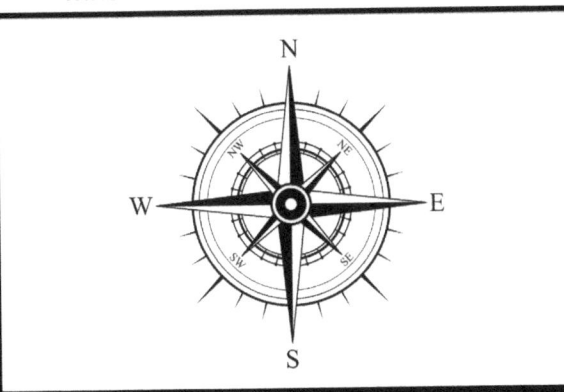

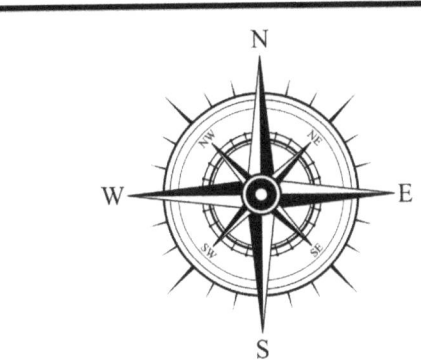

LOCAL WIND SPEED RECORD:

Draw a rough sketch of your property & map out where windbreaks are needed

FROST POCKET ZONES

Mapping frost pockets using seasonal observation or temperature monitoring helps you make smarter plant species and placement decisions, and extends your growing season.

Avoid placing frost-sensitive crops, young fruit trees, or structures in frost pockets. Instead, use these areas for hardy perennials, wetland plants, or water catchment.

Common frostbelt locations include: (Circle all that apply to your property)
- Next to ponds, rivers, or water features,
- North sides of structures or land slopes
- Lowest point of property
- Elevated or raised garden beds with overhead sprinklers
- Open clearings that are lower in elevation
- On/near concrete or stone
- On/near metal structures such as poles, gates, or sculptures
- Bottom of slopes
- OTHER:_____

AVERAGE FROST DATES

| FIRST SPRING FROST DATE:_____ | LAST SPRING FROST DATE:_____ |

FROST DEPTH / FROST LINE

This refers to the maximum depth to which the ground is expected to freeze in a specific location during winter. This is crucial for determining minimum foundation depths to prevent frost heave damage to structures & water catchment systems. Online frost depth calculators and maps can provide estimates, but always cross-reference with local sources for accuracy. For more exact data, refer to local building codes or authorities, which often specify minimum footing depths based on frost line requirements.

FROST LINE DEPTH: _____

COLOR IN THE MONTHS WHERE FROST IS POSSIBLE IN YOUR AREA

JAN	FEB	MAR	APR	MAY	JUN	JUL	AUG	SEP	OCT	NOV	DEC

PROPERTY FROST POCKET MAP

Draw a rough sketch of your property & map out where the frost pockets are. These places are where cold air collects & are frost prone. Frost pockets are unique microclimates that can be used advantageously if identified & utilized correctly.

CALCULATING THE RAINWATER BUDGET

A rainwater budget estimates the total amount of rainwater available on a property. This is necessary to know on properties with no water, and beneficial to know in addition to your other water rights. Rain that falls on impermeable surfaces such as roofs and concrete slides off. Channeling that water helps in designing efficient water management systems for irrigation, storage, and conservation. In order to calculate how much rainwater you can channel from your impermeable surfaces you need to calculate the square footage of all non permeable surface areas including roofs and concrete.

FORUMULA:

Catchment Area in square feet × Rainfall depth × 0.623 = Gallons of Water Collected

Example Calculation:
Catchment area (roof or land surface) = 100 sqft
Annual rainfall = 15 inches
100 x 15 x 0.623 = 934.5 gallons of water collected

TOTAL ANNUAL RAINWATER BUDGET:_____

CALCULATING THE TOTAL WATER BUDGET

Add up the total amount of water you have from the rainwater and any other water rights or shares you may have. That gives you a good idea of what and how much you can grow, store, and distribute throughout the property. If water supply is less than demand, adjust by:

- Adding more catchment areas (rooftop collection, landscape runoff).
- Increasing storage capacity (ponds, cisterns, tanks).
- Using water-efficient irrigation (drip irrigation, ollas, mulching).
- Recycling greywater for irrigation.
- Reducing consumption by planting drought-resistant crops and using water management (swales, keyline design, etc).

SOLAR ASPECT MAP

A Solar Aspect Map is a design tool used to analyze how sunlight moves across a landscape throughout the year. It helps determine the sun-exposed areas and shade patterns in both the winter and the summer. This in turn helps you to position solar panels, greenhouses, and water catchment systems. It helps you identify microclimates & frost pockets for season extension, and assist with passive solar structures, all of which are essential for garden placement, energy efficiency, and overall site planning.

HOW TO CREATE A SOLAR ASPECT MAP USING GOOGLE EARTH

1 Open Google Earth Pro (desktop version, free to download).
Use the search bar to locate your property.
Adjust the zoom level to get a full view of your landscape.

2 Click the Sun Icon (☀) in the toolbar to activate the Sun Time Slider.
Select the solstice dates using the time slider:

Winter Solstice shows the lowest sun angle, highlighting shaded areas and frost pockets.

Summer Solstice shows the highest sun angle, revealing sun-exposed areas and potential overheating spots.

Move the time slider throughout the day (morning, noon, afternoon) on both solstic days to observe sun movement and shadow length throughout the day.

Take screenshots of your land at different times of the day and on both solstices. Print or import the images into a design tool (Google Drawings, Adobe, Procreate, etc) Copy/paste images into a Doc, or hand-sketch on paper.

Mark existing sun and shade aspects. Keep in mind that sun and shade aspects can change as trees or buildings are erected or removed.

- Highlight full-sun areas: Ideal for gardens, orchards, and solar panels.
- Mark partial-shade areas: Best for shade-loving plants, seating areas, or passive cooling strategies.
- Indicate year-round shade: Suitable for structures, compost, or water storage to reduce evaporation.
- Identify wind corridors & heat traps: Adjust design accordingly for windbreaks or cooling strategies.

GOOGLE EARTH
SOLAR ASPECT MAP EXAMPLES

These screenshots show the sun and shade dynamics on two separate properties on the solstices. The top is a screenshot of a property right after sunrise/sunset on the summer solstice, just as light officially touches the property. The bottom screenshot shows a property's sunlight on both solstices during working hours. Different clients, different needs, different research, but both showcasing the swings of daylight to assess for best plant placements.

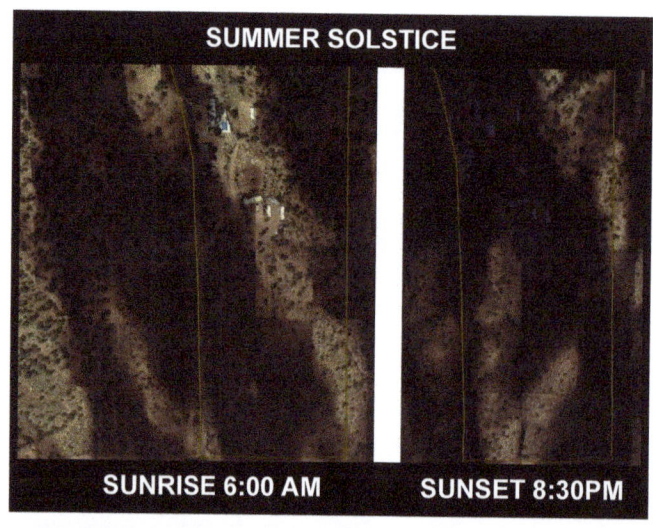

SOLAR ASPECT MAP

Copy/Paste images from Google Earth Desktop using the sun tool. Using the slider, change the dates to the winter and summer solstice to see the sun/shade dynamics throughout the year.

SUMMER

WHAT TIME DOES THE SUN TOUCH THE PROPERTY IN THE MORNING ON THE SUMMER SOLSTICE? _____
WHAT TIME DOES THE SUN SET ON THE PROPERTY IN THE EVENING ON THE SUMMER SOLSTICE? _____
TOTAL HOURS OF SUNLIGHT ON THE SUMMER SOLSTICE: _____

WINTER

WHAT TIME DOES THE SUN TOUCH THE PROPERTY IN THE MORNING ON THE WINTER SOLSTICE? _____
WHAT TIME DOES THE SUN SET ON THE PROPERTY IN THE EVENING ON THE WINTER SOLSTICE? _____
TOTAL HOURS OF SUNLIGHT ON THE WINTER SOLSTICE: _____

64

SOLAR / SHADE ASPECT MAP

The sun rises in the east and sets in the west. This gives east facing land morning sun and afternoon shade, and west facing land the opposite. South facing gets full sun and north facing remains shady. Evaluating your land in this way will help with placing elements where they are best suited.

MOSTLY SHADY
Shade loving plants
Mushrooms
Water storage
Animal pens

PM SUN / AM SHADE
Herbs
Hardy plants
Trees

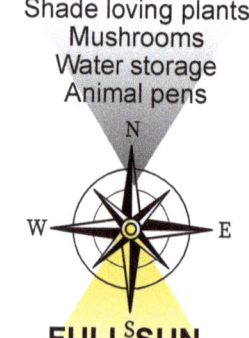

AM SUN / PM SHADE
Sensitive plants
Berries
Gardens in hot
climates

FULL SUN
Gardens in cool climates
Hardy plants
Orchards & Veggies

Draw a rough sketch of your property & map the general shade dynamics on the property

SECTION 4:
GEOLOGY ASSESSMENT

Geology considers land and soil composition in relation to the surrounding landscape. It is not about the life in the soil or what can grow on the land, it is about the structural dynamics of the land itself. It includes gathering soil data and considering the geological influences or nearby natural features that affect the soil structure.

By grounding your design in alignment to the geology your property sits on, you ensure that it fits harmoniously into both the human and natural landscape.

GEOLOGICAL CONCEPTS

ALLUVIUM: Sediment (sand, silt, and clay) deposited by running water, which often creates fertile floodplains and deltas.
RELEVANCE: Alluvium deposits can be found near rain gutters, ditches, and waterways and can be naturally fertile growing spaces.

BEDROCK: The solid, unweathered rock lying beneath the soil.
RELEVANCE: Distance to bedrock indicates how deep water, roots, and nutrients can move through the soil. Shallow bedrock limits infiltration and root growth, affecting where you can place trees, ponds, or earthworks, while deep soils allow for better water storage and stronger plant anchoring.

BIOREGION: A geographic area defined by natural boundaries such as watersheds, climate patterns, landforms, and native species rather than political lines.
RELEVANCE: Helps you design in harmony with the local ecology, using plants, materials, and strategies suited to the natural rhythms, resources, and limitations of your specific environment.

COLLUVIUM: Unconsolidated, unsorted material deposited at the base of slopes by gravity and runoff.
RELEVANCE: Colluvium can be a sign of potential hazards such as rock fall, unstable soil, or steep slopes.

ELECTROCULTURE: The practice of using weak electrical currents or atmospheric energy to stimulate plant growth and soil activity, often through copper antennas or magnetic devices placed in the garden.
RELEVANCE: Enhances natural energy flows, increases soil vitality, and reduces external inputs.

HUMUS: The stable, decomposed organic matter in soil that improves its fertility and structure.
RELEVANCE: Humus is what's created through composting.

LOAM: A balanced mixture of sand, silt, and clay.
RELEVANCE: Loam is considered perfectly balanced soil in regards to water retention and nutrient capacity. Humus mixed with loam is considered the best garden soil.

PARENT MATERIAL: The geological material from which soil is formed such as sandstone, granite, limestone, and basalt.
RELEVANCE: Can affect structures, water retention, and water runoff.

SOIL COMPOSITION: Understanding the sand, silt, and clay ratios and their impact on drainage and fertility.

RELEVANCE: *Guides amendments and crop choices for soil improvement.*

Sand, silt, and clay are the three main mineral components of soil, and their proportions determine a soil's texture, drainage, and nutrient-holding capacity:

 **Sand has the largest particles. It feels gritty and drains quickly, but doesn't hold nutrients or water well.**

 **Silt has medium-sized particles. It has a flour texture when dry and feels smooth and silky when wet. It holds water better than sand, and provides moderate fertility.**

 **Clay has the smallest particles. It feels sticky when wet and hard when dry, holds water and nutrients very well, but can drain slowly and become compacted.**

TYPE / TEXTURE	INCHES OF WATER CAPACITY PER FOOT OF SOIL DEPTH
COURSE SAND	.23 - .75
FINE SAND	.75 - 1.00
LOAMY SAND	1.10 - 1.20
SANDY LOAM	1.25 - 1.40
CLAY	1.20 - 1.40
SILTY CLAY	1.40 - 1.70
FINE SANDY LOAM	1.50 - 2.00
SILTY CLAY LOAM	1.80 - 2.00
SILTY LOAM	2.00 - 2.50

Plants & Soil Sciences eLibrary: *http://passel2.unl.edu/view/lesson/0cff7943f577/10*

SOIL ANALYSIS

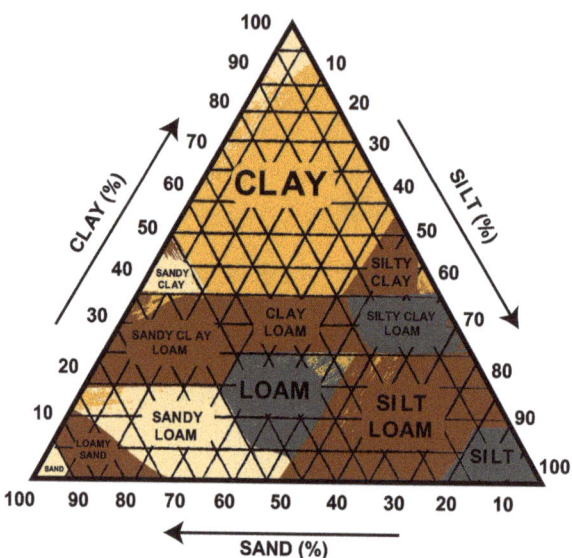

Soil is the easiest thing to fix and change in an ecosystem. Whatever your soil type, a few additions and changes can quickly change undesireable soil into fertile growing soil. In order to do that efficiently you need to know what you are currently working with. This will give you direction on how to amend the soil accordingly.

There are 2 ways to figure out your soil composition: a jar sample, or by researching your local online soil surveys. If you're in the United States you can use the USDA Web Soil Survey website. ***Outside of the United States, research your country's Soil Survey departments to see if an online database is available for your area.***

 **JAR SAMPLE**

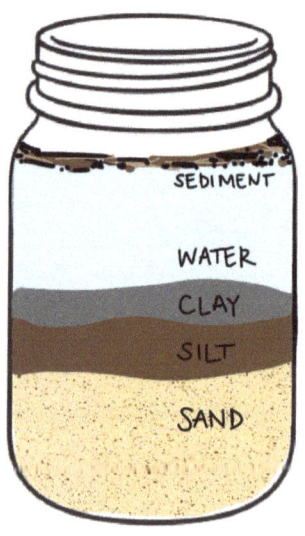

Fill a clear jar halfway with soil, then add water leaving a couple inches of space at the top. Shake the jar well and let it sit undisturbed for 12 hours or overnight.

The soil will settle into layers: sand on the bottom, silt in the middle, and clay on top. Measuring each layer's thickness reveals the soil's texture and the relative percentages of sand, silt, and clay in your sample.

SOILWEB: AN ONLINE SOIL SURVEY BROWSER

Go to: https://casoilresource.lawr.ucdavis.edu/gmap/

Zoom into your property until you can see the yellow lines that dilineate the different soil types.

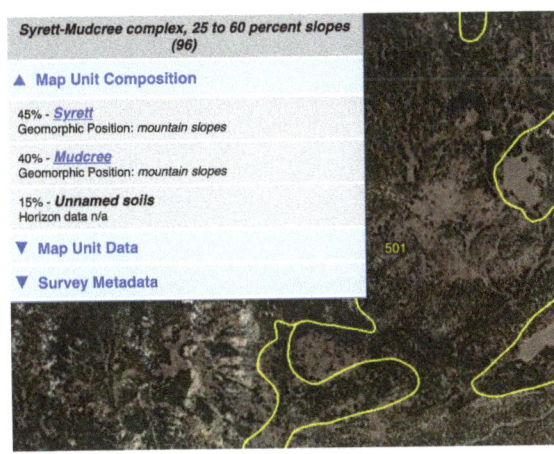

Click anywhere on the map and a little blue box will appear in the left corner of the screen with that area's soil information. You'll want to identify all of the soil types within your property boundaries. It is common to have more than 1 soil type on any given parcel of land.

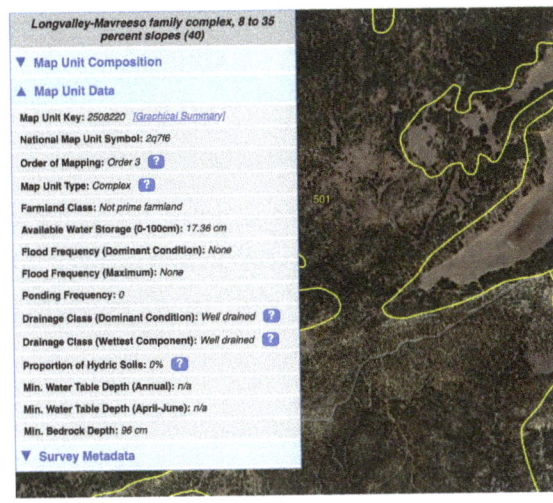

Click the "Map Unit Data" button for a complete list of soil data. Drainage class, Flood Frequency, Water Table Depth, and Bedrock Depth are all important to note.

SOIL ACIDITY JAR TEST RECIPE

Whether soil is alkaline or acidic plays a key role in what type of plants you can plant on your property. Keep in mind that one location on your property may be acidic and another alkaline, it just depends on how much land you are working with and how many micro ecosystems exist within it. To test the acidity of soil at any location on your property you will need:

2 Jars
Baking Soda
Soil
Distilled Water
Vinegar

Fill each jar up with ¼ C of soil. Add ½ cup of distilled water to each cup. Add ½ C vinegar to one jar, and ½ C baking soda to the other jar. If the mixture with the vinegar fizzes & bubbles the soil is alkaline. The more vigorously it fizzes the more alkaline it is. If the mixture with the baking soda fizzes & bubbles the soil is acidic.
The more vigorously it fizzes the more acidic it is.

IS THE SOIL ALKALINE OR ACIDIC? (circle)

pH scale

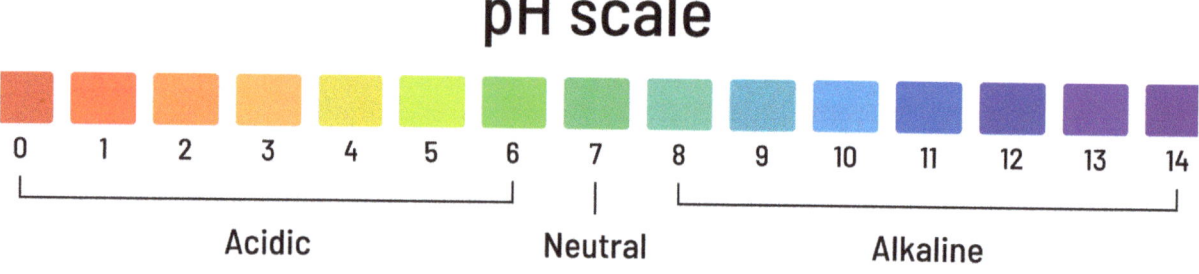

Be sure to test multiple locations on your property

**You can also use soil PH testing strips for the most accurate reading.
They can be found at your local nursery and in many general store
gardening sections**

PH SCALE:_____

GEOLOGY ANALYSIS

Using the USDA Web Soil Survey or your own country's soil survey website, discover the following:

SOIL COMPOSTION (Sandy Clay Loam, Silty Clay Loam, etc):

_____

TYPICAL PROFILE LAYERS:

SURFACE:_____

SUBSURFACE:_____

1+FEET DEEP:_____

3+FEET DEEP:_____

PARENT MATERIAL:_____

DEPTH TO RESTRICTIVE FEATURE:_____ **DEPTH TO WATER TABLE:**_____

DRAINAGE CLASS (circle)

Excessively drained   Somewhat excessively drained   Well drained

Moderately well drained   Somewhat poorly drained

Poorly drained   Very poorly drained

AMMENDMENTS THAT IMPROVE SOIL COMPOSTION		
Sandy soil needs more organic matter to hold water Silty soils need structure and stability Clay soils need aeration and organic matter to improve drainage		
Sand	Biochar Worm castings Compost Manure	Peat moss or coconut coir Leaves Clay-rich subsoil (in small amounts) Green manures and cover crops
Silt	Biochar Gypsum Regular mulching Fine gravel	Fibrous or woody compost Woodchips Deep rooted cover crops
Clay	Woodchips Gypsum Biochar Straw	Course fibrous compost Leaves Deep rooted cover crops Sand (in small amounts, mixed thoroughly

SECTION 5:
TOPOGRAPHY ASSESSMENT

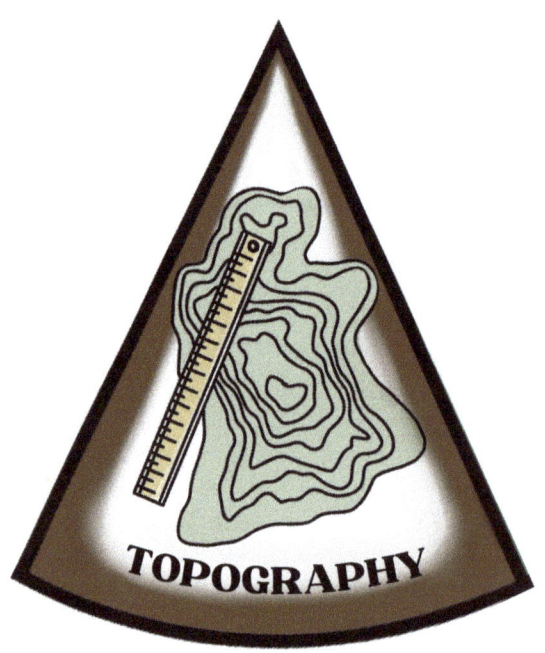

Topography is the study of the physical shape and layout of the land, its slopes, elevations, contours, and drainage pathways. These natural features control how water flows, how soils build up or erode, and where energy is stored or lost across the property.

By carefully observing and mapping topographical features, you can make design decisions that work with the land's natural form rather than against it.

This allows you to place structures, gardens, ponds, and pathways in ways that conserve energy, prevent erosion, and maximize the functionality and beauty of the site.

TOPOGRAPHICAL CONCEPTS

ASPECT: The compass direction that a slope faces.
RELEVANCE: In the Northern Hemisphere, south-facing slopes get more sun (warmer, drier), while north-facing slopes are cooler and retain more moisture. Aspect guides plant placement, microclimate creation, and sun harvesting strategies.

CATCHMENT AREA: The area from which water drains into a specific feature like a pond, swale, or tank.
RELEVANCE: Calculating catchment area helps determine how much water can be harvested and where to place earthworks or storage features.

CONTOUR LINES: Imaginary lines on a map or landscape that connect points of equal elevation.
RELEVANCE: Contour lines guide the layout of swales, berms, paths, and planting beds. Building along contours helps slow, spread, and sink water evenly into the landscape, preventing erosion and maximizing infiltration. You can find contour maps of your area online. They are especially helpful for larger properties with varied topography.

ELEVATION: The height of land above sea level.
RELEVANCE: Affects climate, frost risk, and plant selection. Higher elevations are cooler and may have different water patterns, UV and wind exposure, or planting windows.

RIDGE: A high area of land running along a slope, often dry and exposed.
RELEVANCE: Ridges shed water and are often ideal for access paths, structures, or windbreaks. Understanding ridgelines helps prevent water loss and guides how to contour plantings for wind and sun exposure.

SLOPE: The degree or angle of incline or decline of the land's surface, usually measured as a percentage or ratio.
RELEVANCE: Slope determines how water, soil, and nutrients move across the landscape. Gentle slopes allow for better infiltration and stable plantings, while steep slopes increase runoff and erosion risks. Understanding slope helps guide placement of swales, terraces, access paths, and structures to maximize water retention, prevent soil loss, and ensure long-term landscape stability.

SLOPE BREAKS: Points in the landscape where a slope changes grade or direction.
RELEVANCE: These are excellent locations for placing swales, terraces, check dams, or ponds, as water slows down and infiltrates better at these transition points.

SWALES: Shallow, level-bottomed ditches dug on contour that collect and infiltrate rainwater.
RELEVANCE: Swales are key for rainwater harvesting and passive irrigation. They recharge groundwater, reduce runoff, and can be planted with trees or perennials to create lush, productive edges.

TERRACING: Shaping steep slopes into level or gently sloped steps.
RELEVANCE: Allows for planting, water retention, and erosion control on steep land. Used heavily in drylands and mountainous sites.

WATERSHED: The land area that channels rainfall and snowmelt, ultimately to creeks, rivers, and lakes.
RELEVANCE: Understanding your site's place in a larger watershed helps you manage runoff, prevent downstream impacts, and design in harmony with natural hydrological flows.

IMPORTANT NOTES ABOUT SLOPE

Slope is one of the most critical topographical features in permaculture and regenerative design as it directly affects water movement, erosion risk, accessibility, and planting options. While every site is unique, here are some general guidelines to use when working with sloped land:

A 1% slope means a 1-foot drop over 100 feet of horizontal distance.
A 1% gentle slope allows water to slowly move across the landscape, enabling infiltration without erosion. Best for swales, orchards, gardens, water harvesting, and broadacre farming.

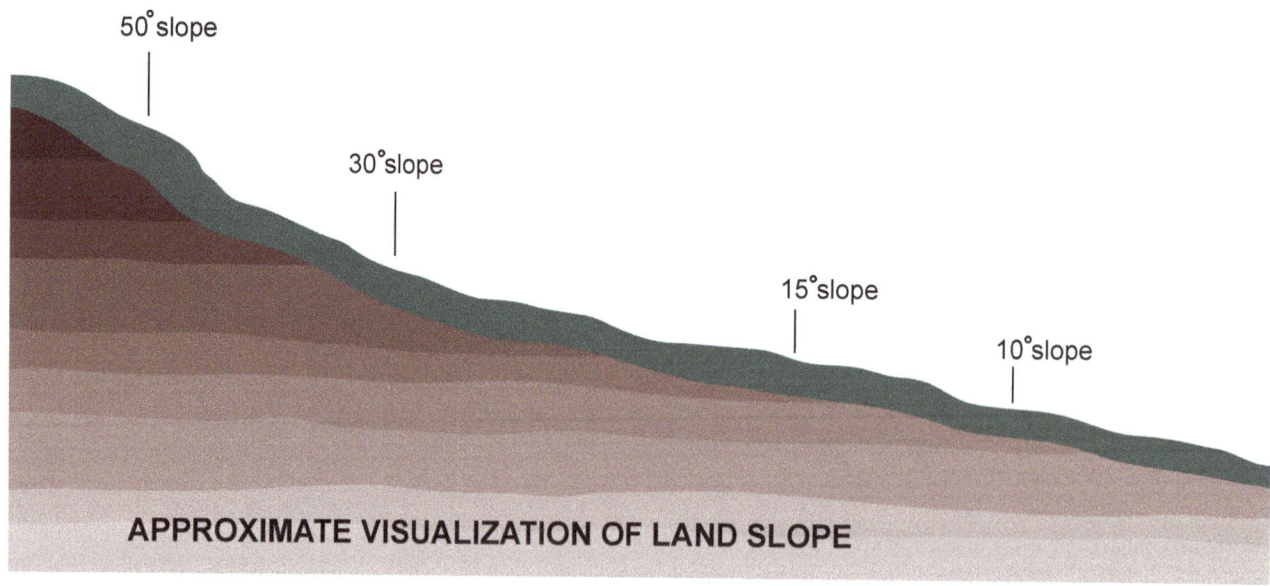

50° slope

30° slope

15° slope

10° slope

APPROXIMATE VISUALIZATION OF LAND SLOPE

 REMEMBER!

- *Your roads should run along contour or gently curve across slope with proper water drainage.*
- *Never build roads straight up or down steep slopes without serious engineering.*
- **Avoid Swales on Slopes > 15%. Swales on steep slopes may fail or cause erosion if not reinforced. Instead, use terraces or leave steep slopes forested and stabilized**

WHEN TO USE TERRACING INSTEAD OF SWALES:

Use terraces when slope > 12–15%, especially if the soil is loose, rainfall is intense, or erosion is evident. Terraces create flat planting areas that retain water and soil. Terraces are more labor-intensive than swales, but offer greater long-term stability on steep land. Slopes above 33% generally require retaining walls or structural reinforcements, and the professional standard is to have slopes greater than 33% professionally evaluated by an engineer before terraces are built.

SWALE SPACING BASED ON SLOPE: Swales are placed on contour and their vertical spacing (drop between swales) depends on the steepness of the slope. A common rule of thumb:

SLOPE (%)	SWALE SPACING (VERTICAL DROP BETWEEN SWALES)
2 - 5%	Every 1.5 - 2 feet of elevation drop
5-12%	Every 2 - 4 feet of elevation drop
12-15%	Every 4 - 6 feet of or more. Consider terracing

Tip: As slope increases, swales must be closer together vertically to manage water safely and prevent erosion.

WHEN WATER HARVESTING USING THE SLOPE:

- **Always start water design at the highest practical point.**
- **Use swales, terraces, and infiltration basins to slow and soak water as it moves downslope. Slow & spread the flow.**
- **Overflow from one swale should ideally be caught by the next one downslope.**

WATER FLOW MAP

A Water Flow Map helps visualize how water moves across the land. This includes rainwater runoff, existing sprinkler or irrigation water, and natural drainage paths. It also highlights erosion-prone areas, and opportunities for water harvesting. Mapping water flow ensures efficient water catchment, soil hydration, and flood prevention while working with natural hydrology.

STEPS TO CREATING A WATER FLOW MAP

(1) **START WITH YOUR BASEMAP**
Use Google Earth, local GIS data, or a topographic map to note the geographical features and slope of the land.
- Note the general slope of the land. Even "flat" land has a subtle drainage slope with a water exit point.
- Mark key features like buildings, roads, slopes, trees, ponds, and existing water sources as they obstruct & divert water flow.

(2) **MARK HOSE SPIGOTS & WATER ENTRY AND EXIT POINTS**
- Mark hose spigots and wells
- Mark entry points – Where water exists and/or enters the property. (rain runoff, creeks, irrigation channels, neighboring land drainage).
- Exit points – Where water leaves the property (streams, ditches, drainage swales).

(3) **IDENTIFY THE LAND'S NATURAL WATER FLOW PATTERNS**
Mark flow paths and note how water moves across the land (rushing downhill, seeping into soil, pooling). Use arrows to denote directional flow. Note any problem areas that are exceptionally dry, wet, or prone to pooling or flooding. Through simple earthworks and mindful designing these problems can be mitigated or even turned into net positive factors.

Once you have a good idea of how much water you have going through your property and how it is all flowing you can use that information to design a great water management system. Refer to the waterflow map as often as necessary and update it as the design unfolds if desired.

EXAMPLES OF WATERFLOW MAPS

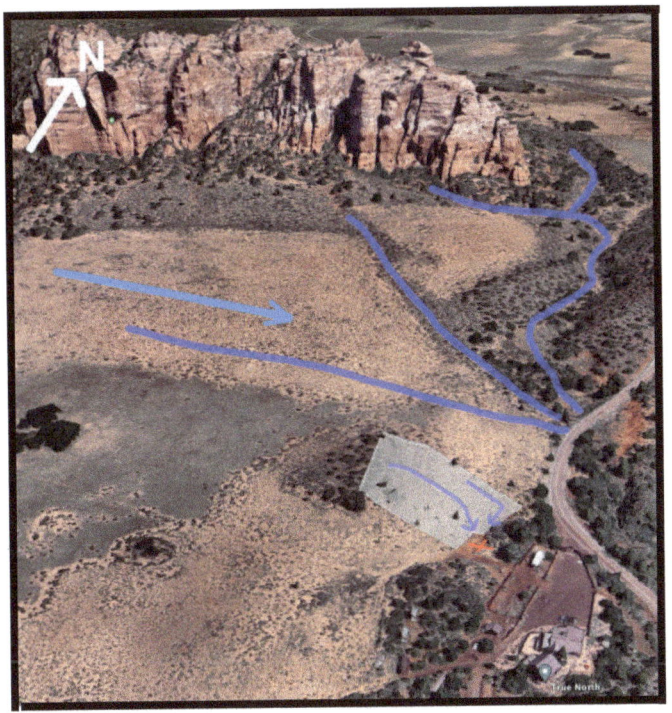

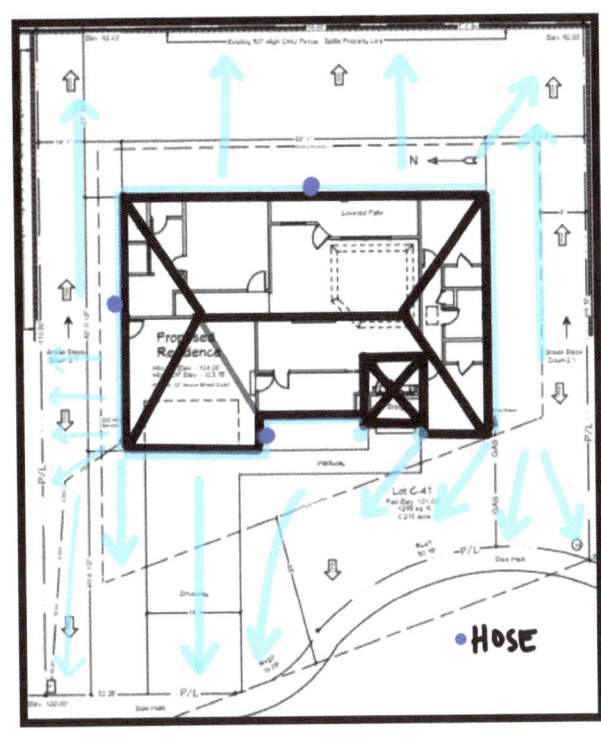

CREATE A WATERFLOW MAP

Using your base map or an image from Google Earth, draw the existing water flow on the property.
Be sure to mark areas of concern such as flood zones and dry spots.

TOPOGRAPHICAL ANALYSIS

ASPECT: This is the direction the property's overall slope is facing. If the property sits on a hill top or on flat land, mark the directions of the slopes surrounding the property. The direction the land is facing is important to consider because it indicates sun exposure. MARK THE DIRECTION ON THE COMPASS.

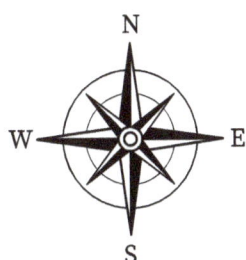

ELEVATION:_____

SLOPE & ELEVATION PROFILE:

To see how your land rises and falls, you can easily create an elevation profile using Google Earth Pro's desktop version.

- **Open Google Earth Pro on your computer and zoom in to your property until you can clearly see your area of interest.**
- **Click the Path Tool (the icon with three connected dots) from the top toolbar.**
- **On your map, click once to mark your starting point and click again to mark your ending point along the path you want to measure such as across your property. A line will appear connecting the two points.**
- **In the Path window that appears, you can name your path if you wish, then click OK to save it.**
- **In the Places panel on the left side of the screen, right-click the new path (it will likely be labeled "Untitled Path" if you didn't name it).**
- **From the menu, select "Show Elevation Profile."**

A graph will appear at the bottom of your screen showing the elevation changes along your chosen path. As you move your mouse along the graph, the corresponding point on the map highlights in red allowing you to visualize the rise and fall of your land in real time.

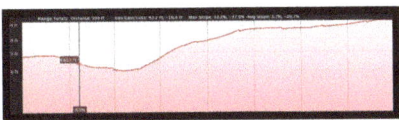

CONTOUR MAP

Contours are imaginary parallel lines on a map that indicate elevation changes. Contours are important to identify in order to design water management systems that slow and spread out the flow of water.

If you are in the United States you can use the USGS TopoView website to identify the contour lines of your property. Otherwise, research your country's available online topographical maps.

SECTION 6:
ECOLOGY ASSESSMENT

Ecology is the study of how natural systems function, focusing on the cycles of water, energy, nutrients, and the relationships between organisms and their environment.

When applied to land design, ecology reveals how to create systems that regenerate rather than deplete resources. Working with ecological principles allows you to reduce waste, conserve water, build soil, and strengthen the natural resilience of your property.

By aligning your design with these cycles, you create landscapes that are not only sustainable but also life-enhancing for generations to come.

ECOLOGICAL CONCEPTS

BIODIVERSITY EQUALS RESILIENCE: If one element or plant type fails, others can fulfill similar roles, keeping the system stable and productive.
RELEVANCE: Just like in nature, a diverse ecosystem can adapt, recover, and continue thriving even when conditions change.

BIODYNAMIC GARDENING: A holistic, organic farming method that views the farm as a self-sufficient, living organism, emphasizing soil health, biodiversity, and aligning activities with lunar cycles and cosmic influences.
RELEVANCE: The soil expands and contracts according to the moon cycles, so this gardening practice aligns the sowing of seeds and harvesting of plants according to that cycle to boost nutrients and enhance resilience. The moon also influences many insect cycles—this practice helps mitigate pest damage by timing plant care with lunar rhythms.

BIOME: A large geographic region on Earth defined by its specific climate, distinct plant and animal life, and the unique ecosystems that exist within it.
RELEVANCE: Mimicking local biomes and their inherent biodiversity helps create stable, self-regulating food production systems.

CARRION EFFECT: The natural process where decaying animals (carrion) return nutrients to the soil, enriching it with nitrogen, phosphorus, and other minerals while supporting scavengers and decomposers like insects, fungi, and microbes.
RELEVANCE: This illustrates the principle that waste equals resource. Natural decay processes build fertility, feed soil life, and sustain ecosystem balance without external inputs.

COMPANION PLANTING: The practice of growing different plants together in ways that benefit each other such as deterring pests, improving soil fertility, attracting pollinators, or providing shade and support.
RELEVANCE: Reduces the need for chemical inputs, enhances biodiversity, and builds resilience within the garden or landscape.

CORES, CORRIDORS, & CARNIVORES: A saying to remind you of the importance of carnivores and their necessary migration corridors between core ecosystems. Healthy corridors prevent incest and extinction within species and maintain balanced predator–prey ecosystems.
RELEVANCE: Remember to create wildlife corridors and foster beneficial predator and pollinator habitats. Research your local wildlife corridors and connect your property to conservation efforts whenever possible.

EDGE EFFECT: Transitions between ecosystems or conditions create zones of increased fertility, moisture, and biodiversity which is critical to regenerative design. These edge zones are some of the most fruitful places to plant, observe, or modify for ecological abundance.
RELEVANCE: Intentionally seed, plant, and utilize your edges! They are natural, low-maintenance, water- and nutrient-collecting zones.

FOREST LAYERS: Forests organize into canopy, understory, shrub, herb, ground cover, root, and climber layers.
RELEVANCE: Mimicking these layers increases pest resilience, biodiversity, and productivity in limited spaces.

INTEGRATE DON'T SEPARATE: Design systems where elements work together in mutually beneficial relationships rather than isolating them. Every element should serve multiple functions and support others, creating balance and resilience.
RELEVANCE: A forest ecosystem embodies this principle. Trees, fungi, animals, and plants all interact through shared nutrients, shade, moisture, and habitat.

INVASIVE PLANTS: Non-native plant species that reproduce rapidly, spread aggressively, and cause harm to the environment, economy, or human health.
RELEVANCE: They outcompete native plants for light, water, and space, harming native ecosystems, causing economic losses, and disrupting natural ecological balance.

KEYSTONE SPECIES: The species that have an outsized influence on ecosystem structure and resilience. Removing them often leads to dramatic changes or collapse of entire ecological communities.
RELEVANCE: Research the keystone plants, animals, and insects in your area and incorporate them into your design wherever possible.

NATIVE PLANTS: Species that evolved and thrive naturally in a specific geographic region, adapting to local climate, soil, and wildlife without human intervention.
RELEVANCE: They provide essential habitat and nourishment for pollinators, birds, and insects; require less water and maintenance once established; and help maintain ecological balance.

OVERPLANTING: The practice of planting more than you need, with the expectation that some loss will occur due to pests or disease.
RELEVANCE: This deliberate strategy ensures a sufficient harvest despite natural losses. Examples include planting extra vegetables of differing varieties or allowing fruit trees to grow taller for wildlife.

POLYCULTURE: Growing diverse plants together to mimic natural ecosystems.
RELEVANCE: Reduces pests, diseases, and competition for resources while promoting ecological harmony.

PREDATOR/PREY PEST BALANCE (aka "WHAT EATS THIS PEST?"): A pest control tactic that involves analyzing natural predators of pests and encouraging their presence on the property. When a pest invasion is present ask yourself, "What eat's this pest?"
RELEVANCE: Predators not only eliminate pests but also disrupt their reproductive cycles. Examples include chickens for insects, cats for mice, and owls for rodents.

SOIL FOOD WEB: Soil health depends on a complex network of organisms like bacteria, fungi, and nematodes. Spraying chemicals kills beneficial life and strips the soil of nutrients.
RELEVANCE: Designing for living soil improves fertility, water retention, and carbon sequestration. Lasagna layering and no-till techniques encourage healthy soil food webs.

THE IMPORTANCE OF HABITAT: A given species' presence on a site depends on its ability to disperse, establish, survive, and reproduce within available habitat.
RELEVANCE: Whether eliminating pests or encouraging pollinators, adjusting habitat is the most efficient way to achieve lasting results.

TREE GUILDS: A polyculture of plants arranged around a central tree that support each other's growth and function that includes nitrogen fixers, dynamic accumulators, repellents, ground covers, and pollinator attractors.
RELEVANCE: Tree guilds create low-maintenance, diverse mini-ecosystems that support fruit and nut production, enhance resilience, and reduce pest pressure.

TROPHIC CASCADE: An ecological phenomenon where changes at the top or bottom of the food chain ripple through all trophic levels, influencing the entire ecosystem.
RELEVANCE: We must consider both top-down and bottom-up influences. Supporting insects, fungi, microbes, producers, and predators is essential for maintaining overall ecosystem health.

WILDLIFE CORRIDORS: Strips of natural habitat connecting larger wilderness areas, allowing animals to move safely between them for feeding, breeding, and migration.
RELEVANCE: Corridors maintain genetic diversity, reduce human-wildlife conflict, and restore ecological balance.
    ***URBAN EXAMPLE:** Planting native trees and plants like milkweed for birds and pollinators to be able to hop between during migration.*
    ***RURAL EXAMPLE:** Leaving wild edges or native plantings for animals to cross through.*

ECOLOGICAL ANALYSIS

Fill out the following pages to do an ecological analysis based on the primary ecological concepts that sustain healthy landscapes & ecosystems

GENERAL LOCAL ECOLOGY

CHECK YOUR BIOME:

- ☐ TROPICAL RAINFOREST
- ☐ TEMPERATE FOREST
- ☐ SAVANNA
- ☐ TAIGA (BOREAL FOREST)

- ☐ DESERT
- ☐ CHAPARRAL
- ☐ TEMPERATE GRASSLAND
- ☐ ARCTIC TUNDRA

PROPERTY ECOREGION/ BIOREGION: Google ; [city name, bioregion]

NAME OF THE FOREST THE PROPERTY IS IN OR NEAREST TO IF ANY:
(This will help you determine your local native plants and wildlife)

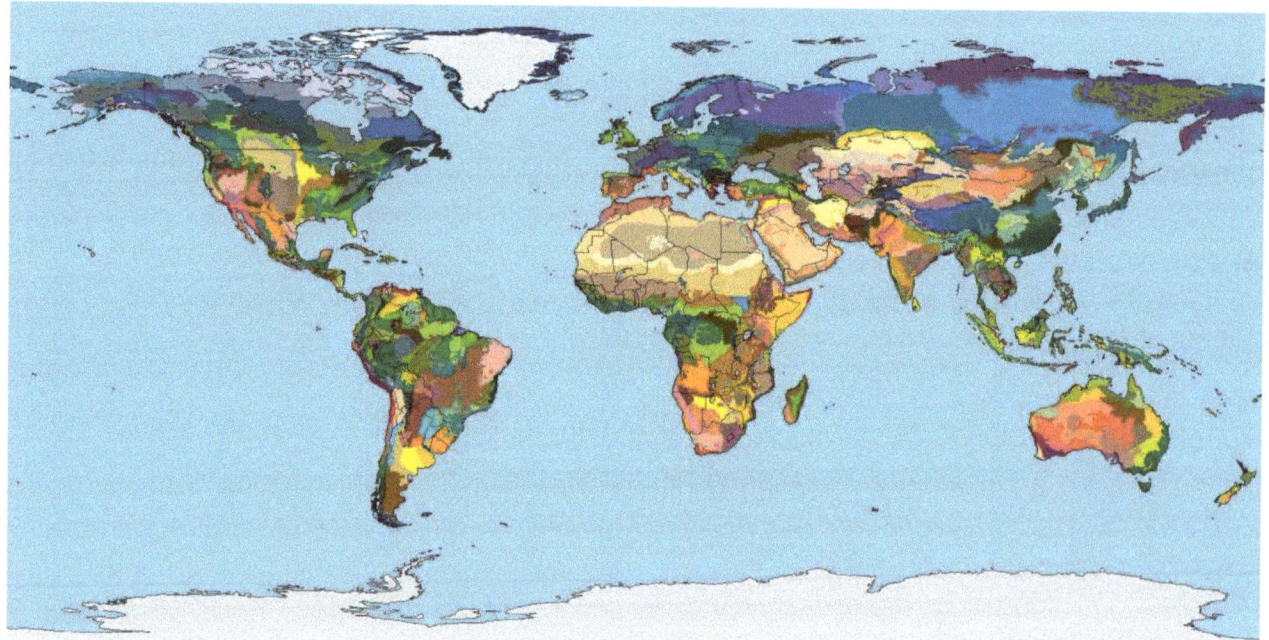

ECOREGIONS OF THE WORLD https://en.wikipedia.org/wiki/Lists_of_ecoregions

FOREST ECOSYSTEM RECIPE

Accounting for the surface area of any given space coupled with the canopy and function of each plant type, generally speaking, you should build your forest ecosystems according to the following guidelines.

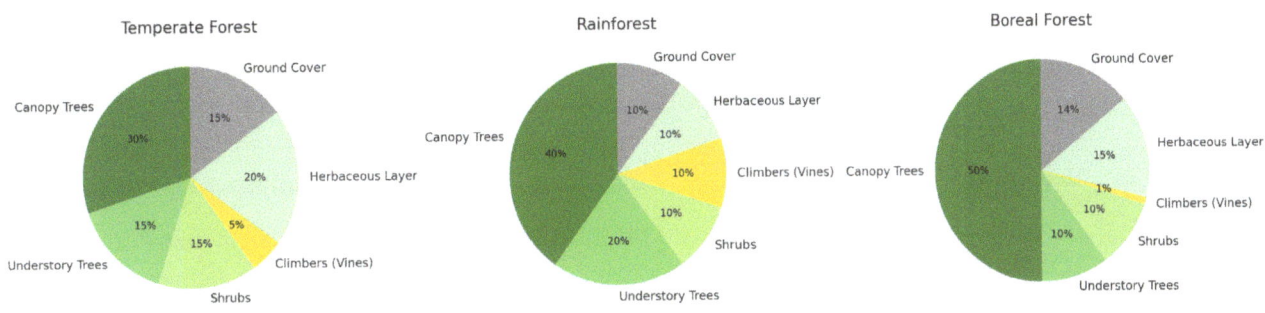

LIST 3 NATIVE PLANTS OF EACH TYPE FROM YOUR AREA		
CANOPY TREES	**UNDERSTORY TREES**	**SHRUBS**
CLIMBERS/VINES	**HERBS/FLOWERS**	**GROUNDCOVERS**
MUSHROOMS		

HALF OF THE PROPERTY SHOULD BE RESERVED FOR NATIVE SPECIES IN ORDER TO UPHOLD THE OVERALL HEALTH & STABILITY OF THE ECOSYSTEM. YOUR PROPERTY IS ONLY AS RESILIENT AS ITS SURROUNDINGS. ENCOURAGE INTEGRATION THROUGH NATIVE PLANT INSTALLATION.

Every forest ecosystem has layers. Each layer supports the others by creating a diverse, stable, and resilient ecosystem, from tall trees hosting owls to tiny ground covers protecting the soil. Together, they form a self-sustaining web of life.

TYPE	FUNCTION
CANOPY TREES	Forms the roof of the forest and provide shelter for birds, insects, and mammals, offers habitat for predatory birds that help control rodent populations, Their leaves create shade, which cools the forest floor and helps retain soil moisture, and drop leaves that become rich mulch, feeding the soil below.
UNDERSTORY TREES	Grow in the filtered light below the canopy, provides habitat and food (fruits, nuts, nesting spots) for smaller birds, mammals, and insects. Acts as a middle layer, buffering wind and adding diversity to the forest. Helps fill in gaps if canopy trees die.
SHRUBS	Offer berries, seeds, and shelter for birds and small animals. Provide nectar for pollinators like bees and butterflies. Create a windbreak and slow water runoff, protecting the soil. Often act as nurse plants for young trees by shading and protecting them.
HERBACEOUS LAYER	Attract and support pollinators and beneficial insects. Feed herbivores, which in turn support beneficial predators. Add beauty and biodiversity to the forest floor. Their decaying leaves build humus and improve soil fertility.
VINES/CLIMBERS	Use trees and shrubs to reach sunlight without needing thick trunks. Provide nectar and shelter for insects, birds, and small mammals. Help connect layers of the forest vertically, acting like bridges for some animals. Increases growing space and adds another food producing layer for humans and wildlife.
GROUNDCOVER (Includes grasses, stonecrops, & all creeping plants)	Blankets the soil to prevent erosion, reduce evaporation, and suppress weeds. Offer habitat for insects, fungi, and decomposers. Help build healthy soil by trapping organic matter and cycling nutrients. Some fix nitrogen or deter pests.
MUSHROOMS	Nature's premier recyclers, breaking down dead organic matter (like leaves, logs, and animals) into fertile soil, releasing vital nutrients (nitrogen, phosphorus) for plants to use, and forming crucial partnerships with plants (mycorrhizae) to share water and minerals, all while serving as food and shelter for wildlife, cycling carbon, and even helping clean up pollution

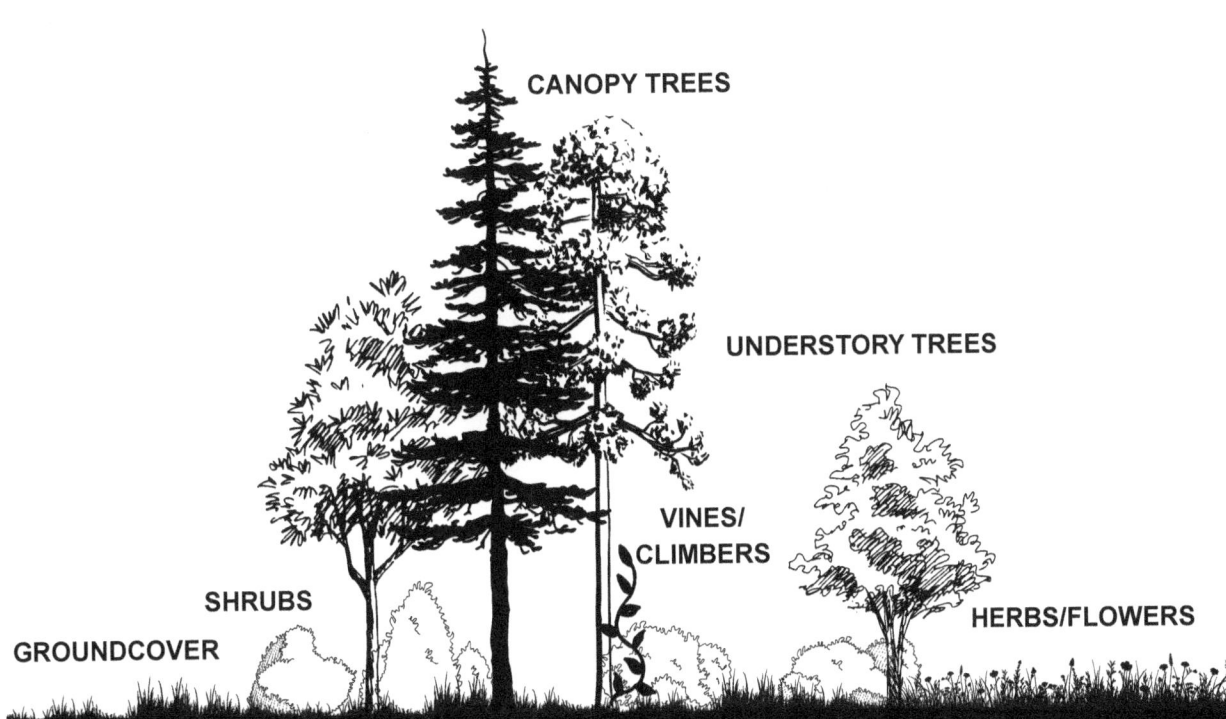

CANOPY TREES

UNDERSTORY TREES

VINES/
CLIMBERS

SHRUBS

GROUNDCOVER

HERBS/FLOWERS

FOREST LAYER PLANTS

DESIGN A FOREST ECOSYSTEM BY SELECTING PLANT SEPCIES FOR EACH LAYER THAT FLOURISH IN YOUR GROW ZONE. WRITE DOWN SEVERAL OPTIONS FOR EACH LAYER

CANOPY TREES	UNDERSTORY TREES (INCLUDES FRUIT TREES)	SHRUBS & BUSHES	VINES & CLIMBERS	HERBS & FLOWERS	GROUND COVERS

NATIVE LOCAL MUSHROOMS SPECIES	OTHER MUSHROOMS SPECIES YOU CAN GROW IN YOUR CLIMATE

FRUIT TREE GUILD

A fruit/nut tree guild is a group of plants that are intentionally planted around a fruit or nut tree to support its health, growth, and productivity. Each plant in the guild serves a specific function to create a self-sustaining mini-ecosystem.

PLANT ROLE	PLANT FUNCTION
NITROGEN FIXERS	Nitrogen Fixers (like clover or peas) add nitrogen to the soil, which helps feed the fruit tree and nearby plants. This reduces the need for fertilizer.
POLLINATOR ATTRACTANTS	Pollinator Attractants like yarrow or bee balm attract bees and butterflies, helping with fruit tree pollination.
REPELLANTS	Pest Repellants like garlic, mint, or marigold repel harmful insects and reduce the need for chemical pest control.
GROUND COVERS	Ground Covers are low-growing plants like strawberries or thyme which cover the soil, keeping in moisture, preventing weeds, and reducing erosion.
DYNAMIC ACCUMULATORS	Dynamic Accumulators are deep-rooted plants like comfrey which pull up nutrients from deep in the soil, making them available to the fruit tree.

GUILD PLANT EXAMPLES FROM EACH TYPE
(Use the internet to discover more types for your climate)

NITROGEN FIXERS	POLLINATOR ATTRACTANTS	REPELLANTS/ SUPPRESSANTS	GROUND COVERS	DYNAMIC ACCUMULATORS
Indigo	Bee Balm	Garlic Chives	Clover	Comfrey
Beans	Hyssop	Onions	Strawberries	Nettle
Peanuts	Echinacea	Sage	Creeping Thyme	Dandelion
Sweet Pea	Lavender	Oregano	Creeping Jenny	Plantain
Chickpea	Yarrow	Thyme	Grasses	Yarrow
Lupine	Rosemary	Nasturtium	Seedums	Sunflowers
Clover	Borage	Red clover	Chamomile	Burdock
Alfalfa	Calendula	Marigolds	Violets	Mullein
Vetch	Goldenrod	Daffodils	Perennial Rye	Borage
Fava Beans	Salvia	Tulips	Yarrow	Alfalfa

FRUIT TREE GUILD

CREATE A LIST OF PLANTS THAT FLOURISH IN YOUR GROW ZONE THAT YOU CAN PLANT AROUND YOUR TREES TO CREATE A SELF-SUSTAINING ECOSYSTEM.
- NATIVE PLANTS FUNCTION BEST -

NITROGEN FIXERS	POLLINATOR ATTRACTANTS	REPELLANTS/ SUPPRESSANTS	GROUND COVERS	DYNAMIC ACCUMULATORS

90

EDGE EFFECT ANALYSIS

Edge zones are *transition areas between elements* that create zones of increased fertility, moisture, and biodiversity. These edge zones are some of the most fruitful places to plant, observe, or modify for ecological abundance. Check the boxes of the Edge Effects that exist on your property.

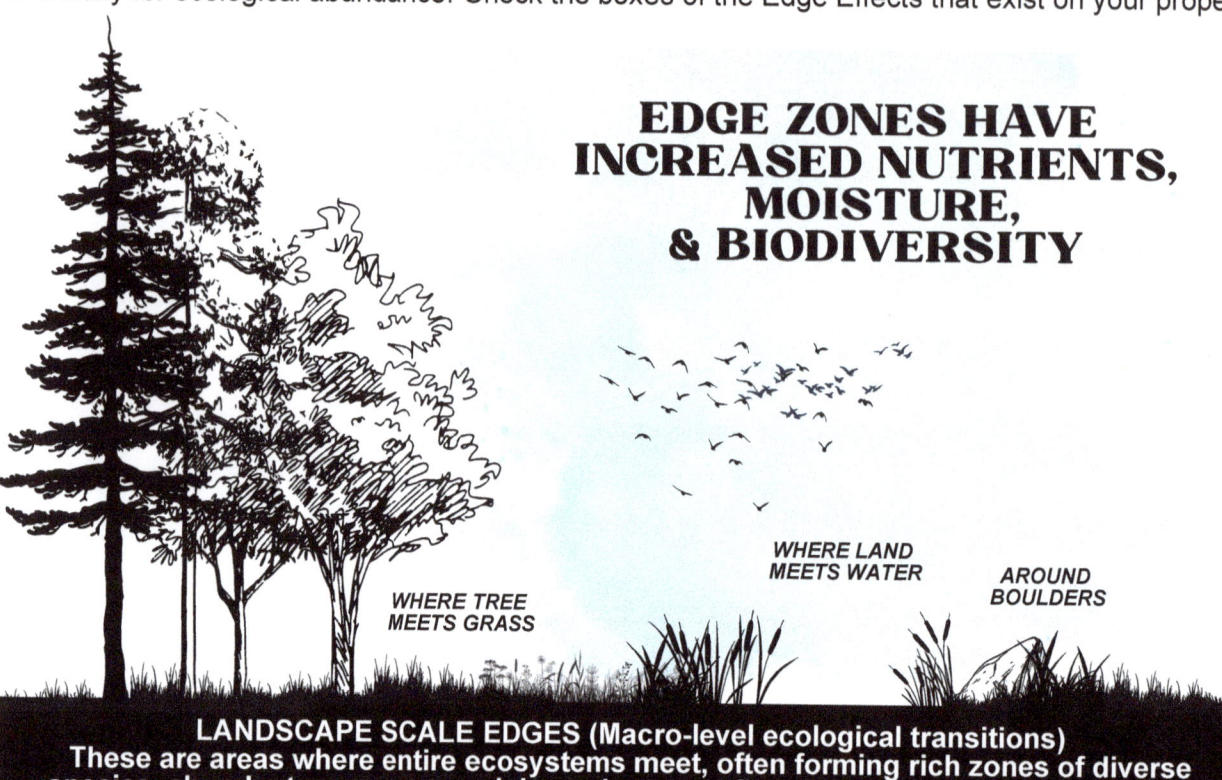

EDGE ZONES HAVE INCREASED NUTRIENTS, MOISTURE, & BIODIVERSITY

WHERE TREE MEETS GRASS

WHERE LAND MEETS WATER

AROUND BOULDERS

LANDSCAPE SCALE EDGES (Macro-level ecological transitions)
These are areas where entire ecosystems meet, often forming rich zones of diverse species, abundant resources, and dynamic energy flow. You can design some of these nutrient rich ecological transition zones into your property with ponds, water features, & mindful plant placement. Check which edge zones you plan to design with.

✓	TYPE	EFFECTS
	FOREST / MEADOW EDGE (Backyard trees & grass)	Grasses, shrubs, and shade tolerant plants coexist. The zone where trees give way to open meadows and grasslands is high in nutrients. Sensitive plants thrive here as do berries and flowers. The zone is high in pollinator and bird activity.
	RIPARIAN ZONE: WHERE WATER & LAND MEET. (Ponds & water features)	Amphibians, reeds, sedges, fish, and aquatic insects abound. Nutrients are high and cycling is rapid. The zone is high in wildlife activity and essential to predator/pest balance.

✓	TYPE	EFFECTS
	FENCE LINES	Wind-sheltered, water-catching, often filled with volunteer plants and vines at the base. Great location for many plants.
	SIDEWALK / GRASS MARGIN	Warmth and runoff make this a weedy, biodiverse strip ideal for opportunistic herbs and great places to plant flowers.
	RAISED BED / PATHWAY EDGE	The edges of pathways and raised beds have high nutrient runoff, irrigation spill, and temperature contrast which encourages rapid growth.
	COMPOST PILE PERIMETER	Soil fertility is highest around compost zones; edge plants flourish with minimal care.
	STACKED MATERIALS	(Wood, rocks, bricks, materials, etc.) Cracks and shade retain moisture allowing for seed germination and insect habitat.
	MULCH / BARE SOIL EDGE	Acts as a fungal frontier where soil organisms thrive and organic matter decomposes quickly.
	POND OR TANK MARGIN	Water availability attracts dragonflies, frogs, reeds, and soil-stabilizing plants.
	ANMAIL PEN EDGES	Nitrogen rich zone, nutrient harvesting zone
	STRUCTURE EDGES	Occupied structures have warm microclimates surrounding the perimeter, especially on the south facing side. North sides can be good for mushroom growth
	TERRACE EDGES	Terraces retain moisture and act as nutrient collection zones allowing for better seed germination and encouraging better growth and plant stability

ENDANGERED PLANT ANALYSIS

Use the internet to identify your local endangered species by typing in your location + endangered species. Foster a relationship with regional endangered species by creating space for their natural habitat or by respecting their habitat in the wild.
List a few endangered plant species in your area in the following boxes.

TREES	SHRUBS/FLOWERS	GRASSES/OTHER

LIST YOUR FAVORITE LOCAL ENDANGERED OR AT RISK PLANT SPECIES HERE:

_____

ARE THERE ANY ENDANGERED OR AT RISK PLANT SPECIES ON OR NEAR THE PROPERTY THAT YOU KNOW OF?

YES   NO

IF SO, LIST SPECIES HERE:

_____

CHOOSE A LOCAL AT RISK OR ENDANGERED SPECIES TO SUPPORT ON OR NEAR THE PROPERTY (Support can include protecting established plants, planting them, or simply through teaching or advocation.)

SPECIES NAME:
_____

WHAT TYPE OF HABITAT DOES THIS PLANT NEED TO SURVIVE?

_____

HOW IS THIS PLANT POLLINATED? _____

PREDATOR / PEST ANALYSIS

One of the most effective ways to manage a pest infestation is to let its natural predators do the work for you. When identifying a pest, ask yourself, "What eats this?" By supporting or introducing those predators through appropriate habitat, access, acquisition, or wildlife corridors, you disrupt the pest's ability to feed and reproduce on your property. Over time, pest populations decline and often move on in search of easier conditions elsewhere. Using the lists below, check all the boxes of the predators that are on or around your property, and check all the boxes of the pests that infest or invade your property.

PREDATOR / PEST ANALYSIS

EXISTING PREDATORS: Carnivorous species that preys upon other species.	EXISTING PESTS: Species that are commonly invasive
LARGE PREDATORS ☐ Bears ☐ Bobcat ☐ Coyote ☐ Dogs (livestock guardians / property protection) ☐ Fox ☐ Mountain Lion / Cougar / Puma ☐ Owl ☐ Predatory birds (Hawk, falcon, eagle, crow) ☐ Wolf **SMALL PREDATORS** ☐ Bats ☐ Ducks ☐ Frogs & Toads ☐ Housecat (Indoor/Outdoor) ☐ Lizards ☐ Opossums (primary predator of ticks) ☐ Raccoons & Skunks ☐ Snakes ☐ Weasels **INSECT ALLIES & ARTHROPOD PREDATORS** ☐ Dragonflies ☐ Lacewings ☐ Lady Bugs ☐ Praying Mantids ☐ Spiders	☐ Ants ☐ Aphids ☐ Bears ☐ Beetles (plant-damaging species) ☐ Cockroaches ☐ Coyotes ☐ Deer ☐ Flies ☐ Foxes / Weasels ☐ Fruit & berry eating birds ☐ Gophers ☐ Grasshoppers ☐ Mice / Rats / Voles ☐ Mites ☐ Mosquitoes ☐ Predatory Birds (when preying on poultry/pets) ☐ Rabbits ☐ Raccoons ☐ Skunks ☐ Snails, Slugs & Grubs ☐ Spiders (when indoors) ☐ Squirrels ☐ Stray Cats / Dogs ☐ Termites ☐ Wasps

PREDATOR BASED PEST MANAGEMENT CHECKLIST

Predator-based pest control works best when multiple predators are supported simultaneously and when pest pressure is reduced through fear, presence, and habitat balance, not eradication.

(Check all that you intend to utilize or apply to your property)

HABITAT & NESTING STRATEGIES
☐ Install owl nesting boxes
☐ Install bat houses
☐ Add raptor perches (hawks, kestrels)
☐ Preserve or install songbird nesting habitat (hedgerows, shrubs)
☐ Retain safe standing dead trees (snags) for cavity nesters

SCENT, PRESENCE & TERRITORIAL DETERRENTS
☐ Apply predator urine (coyote, fox, wolf) along boundaries (Can be purchased online)
☐ Disperse clay-based kitty litter as scent markers
☐ Spread pet fur or dog hair in garden and livestock areas
☐ Rotate scent locations regularly to avoid habituation
☐ Walk property boundaries regularly with dogs

REPTILE & AMPHIBIAN ENCOURAGEMENT
☐ Build rock piles, stone walls, or brush piles
☐ Protect and encourage gopher snake habitat
☐ Add shallow water features for frogs and toads
☐ Eliminate rodenticides and broad-spectrum poisons

GUARDIAN & ACTIVE DETERRENTS
☐ Use livestock guardian dogs or property dogs
☐ Allow ducks to forage in gardens and orchards
☐ Use geese as territorial alarms where appropriate

INSECT PREDATOR SUPPORT
☐ Plant nectar and pollen plants for beneficial insects
☐ Leave leaf litter and mulch for ground predators
☐ Install insect hotels
☐ Avoid chemical sprays that harm beneficial insects

LANDSCAPE DESIGN & FLOW
☐ Create wildlife corridors across the property
☐ Design and preserve edge habitats (forest/meadow, pond/land)
☐ Use fencing strategically to guide predator movement

COMMON PESTS & THEIR PREDATORS

Ants	Ground beetles, birds, (woodpeckers, songbirds), lizards
Aphids	Ladybugs, lacewings, parasitic wasps, nectar plantings
Cockroaches	Lizards, frogs & toads, spiders, ground beetles
Coyotes	Dogs, llamas, human presence, & scent marking
Deer	Predator urine (coyote/wolf), dog scent
Flies	Ducks, frogs & toads, dragonflies, parasitic wasps
Fruit Eating Birds	Raptor perches, owl boxes, nearby predator habitat
Gophers	Gopher snakes, barn owls, hawks, dogs
Grubs	Ground beetles, birds, chickens
Mice / Rats / Voles	Owl boxes, raptor perches, gopher snakes, brush piles
Mosquitoes	Bat houses, dragonflies, frogs & toads, fish in ponds
Rabbits	Predator scent, dogs, raptors, snakes (juveniles)
Raccoons, skunks	Dogs, coyote scent, predator urine
Slugs & Snails	Ducks, ground beetles, frogs & toads, leaf litter habitat
Spiders (indoors)	Outdoor predator balance; reduce insect prey
Squirrels	Hawks & owls, dogs, rotating predator scents

BEST PREDATORS FOR PROPERTY PESTS

LIST THE PREDATORS YOU INTEND TO WORK WITH TO HELP MITIGATE PROPERTY PESTS, AND THEIR PRIMARY HABITAT AND PLANT PREFERENCES. THESE WILL BE PLANTS THAT YOU WILL WANT TO DESIGN WITH

PREDATOR	PLANT PREFERENCES FOR HABITAT OR FOOD

ENDANGERED & AT RISK WILDLIFE ANALYSIS

LIST LOCAL ENDANGERED SPECIES:
Use the internet to identify your local endangered species by typing in your state or ecoregion + endangered species. Every property should be fostering a relationship with its regional endangered species by providing native food and/or native habitat.

BIRDS	MAMMALS	AMPHIBIANS & REPTILES	INSECTS	FISH

LOCAL ENDANGERED SPECIES TO SUPPORT ON THE PROPERTY:

SPECIES NAME:_____

HOW I WILL SUPPORT THIS SPECIES:

☐ HABITAT CREATION   ☐ PLANTING ITS FOOD SOURCE   ☐ ADVOCACY

ELIMINATE	ENHANCE	ESTABLISH	PROTECT
List invasive species or plants that threaten the endangered species you chose	List ways you can enhance your property to encourage protection of the species	List plants you can establish that act as food or habitat for the endangered species you chose.	List ways you can protect the endangered species you chose. (habitat, advocacy, etc)

KEYSTONE SPECIES REVIEW

A keystone species is a plant, animal, or other organism that has a disproportionately large influence on the structure, diversity, and function of an ecosystem relative to its abundance. Even if it is not the most numerous species, its presence (or absence) shapes food webs, regulates populations, and maintains ecological balance. Keystone species often create or modify habitat, control herbivore or prey populations, or provide essential resources that many other organisms depend on. When a keystone species is removed, ecosystems can rapidly unravel, biodiversity declines, invasive species may spread, and ecological processes such as nutrient cycling or water flow can shift dramatically. Because of this outsized impact, protecting keystone species is critical for maintaining long term resilient communities. Common keystone species include:

SPECIES	TROPHIC CASCADE EFFECT	CAUSE OF DECLINE
BEAVER	Creates wetlands increasing wildlife, regulates water flow, increases biodiversity, raises water tables, mitigates erosion and flooding damage. Reduces wildfire risk.	Trapping, habitat loss, river channelization, and dam construction.
BISON	Shapes vegetation, builds soil, supports grassland biodiversity, sustains native seed dispersal. Bison graze quickly and don't naturally hover and linger near water leaving land and riparian areas more intact than cattle.	Overhunting, replacement by livestock, fencing, and land conversion to agriculture.
COYOTE	Controls rodent populations, reduces rodent borne diseases, maintains sensitive ecosystem balance.	Urban development, ignorant predator control programs, and habitat fragmentation.
GOPHER TORTOISE	Creates burrows that support 350 other species, supports fire-adapted ecosystems creating land resilience.	Urban development, habitat loss, fire supression
GRAY & MEXICAN WOLF	Controls ungulate (animals with hooves) populations, enables crucial forest and riparian recovery which stabilizes soils, improves water quality, supports beavers, songbirds, fish, and amphibians, and even reshapes river channels over time.	Extermination programs lack of wildlife corridors, and habitat loss.
INSECTS IN GENERAL	Provide food for birds, amphibians, reptiles, and mammals. Essential to soil formation, decomposition, and nutrient cycling.	Pesticides, habitat loss, pollution, climate change, and artificial light.

SPECIES	TROPHIC CASCADE EFFECT	CAUSE OF DECLINE
PINYON JAY	Pinyon pines are keystone trees, and the jay is the keystone species that maintains them. Indirectly, the pinyon jay affects hundreds of species across the woodland biome because its seed dispersal shapes where forests grow	Habitat loss
SALMON	Salmon, particularly Pacific salmon species, are one of the most powerful keystone species on Earth because they connect ocean, river, and forest ecosystems. Their influence is so widespread that well over 130 species directly depend on salmon for food, nutrients, or habitat structure, and hundreds more benefit indirectly.	Dams, overfishing, water pollution, and rising river temperatures.
MOUNTAIN LION COUGAR PUMA	Especially important when it comes to managing chronic wasting disease (CWD) in deer and elk populations. They selectively target sick, weak, injured, or old animals removing infected deer before they spread it through the herd more keeping deer populations healthier and less densely packed. When mountain lions are removed or heavily persecuted, deer populations often surge, disease spreads more rapidly, vegetation declines, and ecosystems lose resilience, highlighting how crucial these predators are for both ecological balance and wildlife health.	Ignorant no-permit-required hunting laws, habitat fragmentation, and lack of wildlife corridors.
POLLINAT-ORS	Bees, butterflies, moths, bats, and other crucial pollinators enable fruit and seed production in countless plants. Their loss disrupts food chains for herbivores, frugivores, and seed dispersers which can collapse entire ecosystems.	Pesticides (especially neonicotinoids), monoculture farming, habitat loss, and diseases like varroa mites.
PHYTO-PLANKTON	Base of the ocean food web. Support zooplankton, fish, whales, seabirds, and indirectly affects global oxygen and carbon cycles.	Ocean warming, acidification, nutrient imbalances, underwater military testing, and pollution.
PRAIRIE DOGS	Keystone herbivores and burrow engineers that support burrowing owls, black-footed ferrets, foxes, insects.	Poisoning, habitat disruptions, and diseases (plague) exacerbated by cattle ranching

KEYSTONE SPECIES ANALYSIS

Keystone species create ecosystems that are resource abundant with balanced predator/pest systems in place making them imperative to healthy flourishing ecosystems. This reduces time, money, and effort spent on pest control and amendment dispersal. Identifying the keystone species of your local native ecosystem will help you to choose plants that you can work with to restore balance and increase abundance. *Remember: Not all native animals are keystone.* (Find your local keystone species with a quick internet search. Type in your location + keystone species to find the exact species type and name unique to your area.)

MAMMALS	BIRDS	REPTILES
AMPHIBIANS	FISH/SHELLFISH	INSECTS

CHOOSE AT LEAST ONE LOCAL KEYSTONE SPECIES FROM YOUR AREA TO WORK WITH ON YOUR PROPERTY. IDENTIFY THAT SPECIES' PLANT AND HABITAT PREFERENCES AND LIST THEM HERE:

KEYSTONE SPECIES:

_____

PLANT & HABITAT PREFERENCES

_____

KEYSTONE PLANT SPECIES ANLYSIS

Keystone plants foster the food and habitat necessary for wildlife. They are the foundation for all life to sustain itself within the ecosystem. Native plants play the same role, but keystone native plants are so foundational that without them the entire localized ecosystem would collapse. Identify the local keystone plant species within your ecoregion so you can incorporate them into your design.

Remember: Not all native plants are keystone. Be sure to specify and clarify your search results.

GRASSES	SHRUBS/FLOWERS	TREES

KEYSTONE SPECIES PROFILE

Common Name: _____

Scientific Name: _____

Ecosystem Role:
☐ **Predator**   ☐ **Seed disperser**   ☐ **Soil builder**   ☐ **Water engineer**
☐ **Pollinator**   ☐ **Keystone plant/tree** ☐ **Structural modifier**

What do they influence? (Check all that apply):
☐ **Water flow**   ☐ **Vegetation patterns**   ☐ **Soil health**   ☐ **Animal shelter**
☐ **Nutrient cycling**   ☐ **Pollination**   ☐ **Insect control**

RESTORATION ACTION
What could be done on or near your property to support this species?
☐ Foster the habitat of the species with plants the species prefers
☐ Create or protect water access
☐ Connectivity: Create corridors, improve access with mindful fencing
☐ Public education: Advocate for the species
☐ Policy/Conservation action: Support, follow, and donate to protection causes
☐ Other: _____

SECTION 7:
BIOLOGY ASSESSMENT

Biology focuses on the living organisms that make up your property, from plants and animals to insects, pollinators, and the countless microorganisms that bring the soil to life.

Each organism contributes to the overall health of the ecosystem, whether by cycling nutrients, preventing pests, or increasing fertility.

Designing with biological systems in mind means creating landscapes that support biodiversity and balance, resulting in gardens and farms that are self-sustaining, regenerative, and more productive over time.

BIOLOGICAL CONCEPTS

ALLELOPATHY: Some plants release chemicals that inhibit the growth of others.
RELEVANCE: Influences crop placement to avoid negative interactions. For example, sunflowers are allelopathic and can suppress or kill sensitive plants nearby.

BIOCHAR: Charred organic matter (carbonized at high temperatures) that is porous and stable in soil.
RELEVANCE: Enhances soil structure, holds moisture and nutrients, and boosts microbial habitats. Excellent for improving dry, degraded soils.

BIOINDICATOR PLANTS: Also called biomonitors, these are plants or organisms used to detect and assess environmental health by showing visible changes in growth, behavior, or appearance due to pollution or other disturbances. Examples include lichens and algae that absorb atmospheric toxins, and certain weeds that reveal soil fertility or imbalance.
RELEVANCE: Helps you "read the land" and diagnose soil conditions, contamination, or fertility levels within your property's ecosystem.

CHOP & DROP: A technique where pruned branches, leaves, or harvested plants are chopped and dropped directly onto the soil as mulch.
RELEVANCE: Mimics the forest floor by feeding soil microbes, suppressing weeds, conserving moisture, and cycling nutrients in place saving time and energy.

COMPOSTING: The decomposition of organic matter like food scraps, leaves, and manure into humus-rich soil through aerobic or anaerobic microbial activity.
RELEVANCE: Returns nutrients to the soil, reduces waste, and builds microbial diversity that supports plant and soil health. Compost feeds the soil food web, which is key to fertility.

COMPOST TEA: A liquid fertilizer made by steeping compost or weeds in water to extract beneficial microorganisms and soluble nutrients.
RELEVANCE: Provides a free, nutrient-dense amendment to apply to plants.

COPPICING: A traditional woodland management practice of cutting trees and shrubs to ground level, encouraging rapid regrowth of new shoots from the base.
RELEVANCE: Provides a renewable source of timber and fuel while enhancing biodiversity by creating varied habitats and increasing light in woodlands. Common species like hazel and chestnut are coppiced on cycles to produce straight poles for building, firewood, and basketry.

LASAGNE LAYERING: A no-dig, no-till sheet mulching technique that creates nutrient-rich soil by stacking organic materials in alternating layers of carbon-rich "browns" (newspaper, cardboard, leaves) and nitrogen-rich "greens" (grass clippings, kitchen scraps).
RELEVANCE: Over time, these layers decompose into fertile soil, building a raised garden bed from the bottom up.

MYCORRHIZAE (FUNGAL NETWORKS): Symbiotic fungi that form connections with plant roots, extending the root system and helping plants access water and nutrients, especially phosphorus.
RELEVANCE: Supports plant immunity, drought resistance, and nutrient cycling. Minimizing soil disturbance helps preserve these beneficial underground networks.

NITROGEN FIXATION: Certain plants (e.g., legumes) host bacteria that convert atmospheric nitrogen into soil nutrients.
RELEVANCE: Essential for building soil fertility and reducing the need for synthetic fertilizers.

NO-TILL: A method that avoids disturbing the soil by forgoing tilling or digging, instead using layers of organic materials, mulch, and cover crops to build healthy, fertile soil and improve its structure and biology.
RELEVANCE: Mimics natural soil ecosystems, reducing water loss, suppressing weeds, conserving soil carbon, and supporting beneficial microorganisms.

VERMICULTURE: The cultivation of worms (usually red wigglers) to decompose organic matter and create nutrient-rich castings (worm compost).
RELEVANCE: Produces high-quality compost, boosts microbial life, and provides a low-odor way to process kitchen scraps even indoors or in small spaces.

TIP: You can vermicompost directly in the garden using a terra cotta pot. Dig a hole the size of the pot, place the pot in the hole, fill it with worm-friendly food scraps and dried weeds, and cover it with the clay dish as a lid. This integrates the worm bin directly into the garden rather than maintaining a separate system. An excellent example of "integrate, don't separate."

BIOLOGICAL ANALYSIS

BIOINDICATOR PLANTS BY SOIL TYPE *HOW TO READ THE WEEDS TO TELL WHAT THE SOIL IS DOING*	
DISTURBED OR OVERGRAZED AREAS	Mullein, Lamb's Quarters, Oxeye Daisy, Shepherd's Purse, Thistle
ALKALINE OR NEUTRAL SOIL	Shepherd's Purse, Yarrow, Clover
DRY OR DROUGHT-PRONE SOIL	Yarrow, Mullein, Purslane, Ragweed
ACIDIC SOIL	Dock, Buttercup, Moss, Sorrel
HIGH FERTILITY OR NITROGEN-RICH SOIL	Nettle, Lamb's Quarters, Chickweed, Clover
LOW FERTILITY OR DEPLETED SOIL	Yarrow, Oxeye Daisy, Thistle, Ragweed, Mullein
POORLY DRAINED OR WATERLOGGED SOIL	Buttercup, Smartweed, Plantain, Dock, Moss, Nettle
COMPACTED SOIL	Dandelion, Plantain, Dock, Thistle, Bindweed, Ragweed, Purslane

LIST THE WEEDS ON THE PROPERTY & WHAT THEY INDICATE ABOUT THE SOIL.

You may need to use the internet or a plant identification app to identify the weeds and figure out what it indicates about the soil.

WEEDS	LOCATION ON PROPERTY	WHAT IT INDICATES ABOUT THE SOIL

DESIGN ORDER & CONCLUSION

Now that you've finished analyzing and gathering data on all the various components that make up your property, you may be wondering how exactly to take that data and extrapolate it into a cohesive design on paper. Putting the design onto paper isn't complicated, but there is an order that helps. My personal design order is as follows:

DESIGN ORDER

1. ACQUIRE PROPERTY MAPS & GENERATE A BASE MAP
2. RESEARCH EXISTING DATA
3. IDENTIFY THE ZONES OF USE AND DEFINE THE ELEMENTS WITHIN THEM
4. MARK FUTURE STRUCTURES & ELEMENTS
5. DEFINE FENCING/PROPERTY BOUNDARIES & DESIGN EDGE ZONES
6. DEFINE PATHWAYS IN AND BETWEEN THE ZONES OF USE
7. DESIGN EARTHWORKS & WATER FLOW MANAGEMENT
8. DESIGN EVERGREEN LAYER: PLACE EVERGREEN TREES & SHRUBS
9. DESIGN DECIDUOUS TREE LAYER: PLACE CANOPY & UNDERSTORY TREES
10. ASSIGN TREE SPECIES & SPREAD OUT THE HARVEST
11. PLACE SHRUBS, GRAPES/VINEYARDS, & BERRY BUSHES
12. BUILD GUILDS UNDER & AROUND TREES & SHRUBS
13. ASSESS FOR MICROCLIMATES & MAKE ADJUSTMENTS IF NECESSARY
14. PLACE & DESIGN GARDEN BEDS, GROW SPACES, & HERB SPIRALS
15. ADD OPTIONAL AND ADAPTIVE ELEMENTS

Using your basemap and the data you've gathered, begin putting elements throughout the property in order of this checklist from top to bottom. Doing it in this order ensures that the more permanent elements are placed first which ensures cohesion throughout.

You've probably also uncovered some concerns or issues that require some creative solutions, or you may be wondering how to put your design into practice. There are a myriad of creative natural design solutions and maintenance practices, so many in fact that I made another book for it!

"Permaculture & Regenerative Landscape Design Handbook: Design Solutions & Reference charts" (COMING SPRING 2026!)

That book goes into detail about the design order step by step offering information to help you decide on plant types and placement. It also includes creative natural solutions for most design challenges, indigenous and natural gardening techniques from around the world, as well as dozens of charts and diagrams to help you make informed decisions.

In the meantime, if you'd like even more worksheets and guidance to help you understand your land, I invite you to explore my website, Feffylane.com where I have more free printables including a harvest & bloom fill in the blank calendar and an herb garden planner organized by organ system.

ABOUT THE AUTHOR

I began my permaculture journey in 2012 while looking for ways to garden more efficiently, eventually earning my Permaculture Design Certification in 2016 through Oregon State University. The following year, I attended their Professional Designer course and began designing professionally for my local community, eventually specializing in drought-resilient and regenerative landscape design. I now teach others how to design regenerative landscapes both online and within my local community.